# HANDBOOK OF
# ELECTRONIC
# SAFETY PROCEDURES

# HANDBOOK OF
# ELECTRONIC
# SAFETY PROCEDURES

**EDWARD A. LACY**

*Technical Writer*

54172

PRENTICE-HALL INC., Englewood Cliffs, New Jersey

*Library of Congress Cataloging in Publication Data*

LACY, EDWARD A    1935–
    Handbook of electronic safety procedures.

    Bibliography: p.
    1. Electronics—Safety measures.   2. Electronic
industries—Safety measures.   I. Title
TK7835.L33        621.381′028        75-44230
ISBN   0-13-377341-8

10  9  8  7  6  5  4  3  2  1

Printed in the United States of America

PRENTICE-HALL INTERNATIONAL, INC., *London*
PRENTICE-HALL OF AUSTRALIA, PTY. LTD., *Sydney*
PRENTICE-HALL OF CANADA, LTD., *Toronto*
PRENTICE-HALL OF INDIA PRIVATE LIMITED, *New Delhi*
PRENTICE-HALL OF JAPAN, INC., *Tokyo*
PRENTICE-HALL OF SOUTHEAST ASIA PTE. LTD., *Singapore*

To my wife RITA

# CONTENTS

ACKNOWLEDGEMENTS     xi

PREFACE     xiii

1   INTRODUCTION     ɪ

Occupational Safety and Health Act,    *2*
General Rules for Safety,    *5*
Guide to this Book,    *7*
Disclaimer,    *7*

2   ELECTRIC SHOCK: UNSAFE AT ANY VOLTAGE     9

Introduction,    *9*
Recognizing Shock Hazards,    *14*
Testing for Electric Shock Hazards,    *18*
Shock Precautions,    *22*
Fundamental First Aid,    *45*

## 3 STATIC ELECTRICITY 50

Generation, *53*
Detection and Measurement, *55*
Control, *57*
Special Problems, *64*

## 4 LIGHTNING 69

General, *75*
Fundamental Protection Measures, *78*
Antennas and Supporting Structures, *81*
Coaxial Lines and Waveguides, *88*
Protection of Station Building, Personnel,
and Equipment, *88*
Protection of Land Communication Facilities, *90*
Protection of Power Services and Utilization Equipment, *94*
CATV Systems, *97*

## 5 RADIATION HAZARDS 98

Introduction, *98*
Biological Effects, *101*
Electroexplosive Devices, *107*
Color Television, *109*
Klystrons and Hydrogen Thyratrons, *114*
Cold-Cathode Gas Discharge Tubes, *118*
Microwave Ovens, *120*
Radar, *128*
Broadcast and Communications Transmitters, *134*

## 6 LASERS 138

Properties of Laser Light, *142*
Biological Effects of Laser Light, *143*

Safety Precautions,   *148*
Proposed Laser Performance Standard,   *155*


7   LADDERS, TOWERS, AND OTHER AERIAL
    HAZARDS                                              163

   Ladders,   *167*
   Towers,   *172*
   Pole Climbing,   *175*
   Antenna Installation on Roofs,   *178*


8   UNDERGROUND HAZARDS                                  179

   Underground Compacting Auger,   *182*
   Line-laying Machines,   *185*


9   TOXIC AND EXPLOSIVE CHEMICALS                        188

   Aerosols,   *189*
   Batteries,   *190*
   Cadmium,   *192*
   Hazardous Locations,   *192*
   Liquid Crystals,   *192*
   Ozone,   *192*
   Printed Circuits,   *193*
   Subsurface Gases,   *193*
   Trichloroethylene,   *194*
   Waveguides,   *194*


10  NOISE                                                196

   Effects of Noise on Man,   *201*
   Noise Laws,   *204*

Noise Measurement,  *207*
Noise Control,  *209*
Ear (Hearing) Protection,  *217*
Ultrasonics,  *221*
Sonar,  *221*


**11  TOOLS**                                                       223

Special Tool Problems,  *228*
Power Tools,  *232*


**12  PRODUCT SAFETY**                                             235

Consumer Product Safety Act—Public Law 92–573,  *237*
Repair,  *242*


**APPENDIX**                                                        244

First Aid,  *244*
Lifting,  *245*
Fire Protection,  *249*
RF-Energy Burns,  *251*
Cathode-Ray Tubes,  *252*
Underwriters Laboratories (UL),  *253*
Cable Television,  *253*
Emergency Lighting,  *254*
Signs,  *255*
Recommended Reading,  *255*
Addresses,  *256*


**INDEX**                                                           263

# ACKNOWLEDGMENTS

Numerous individuals associated with many companies and government agencies have graciously provided information and illustrations for this book. Their assistance has made this book possible and I extend to everyone my sincere thanks. Appropriate credit has been given throughout the book to the particular individuals, companies, or agencies involved.

Most of all, my thanks to my wife, Rita, who did a great deal in preparing the manuscript and helped in many ways in the completion of this book.

E.A.L.

# PREFACE

With the many advances taking place in the field of electronics, the average technician or engineer may find it difficult or even impossible to keep up with the books, magazine articles, and reports describing and explaining these advances. There is simply not enough time to read material in *related* fields such as safety. However, ignorance of the latest safety hazards and precautions can be disastrous in terms of injury or death to the unsuspecting electronics worker or hobbyist.

As electronics inventions have proliferated, so have the possible hazards. It's no longer possible to summarize electronic safety with the simple admonition of "keep your hands off the high voltage."

Present day hazards are not always obvious. What might be considered an insignificant hazard can be a major problem when cumulative effects are considered. And some "safe" practices of recent years are now being found to be "unsafe" as safety experts learn more about the human body's reaction to chemicals, radiation, etc.

Two basic types of hazards are discussed in this handbook: (1) those caused by electronic equipment, such as high voltage and radiation, and (2) those that might conceivably be encountered by an electronics technician or engineer in his work, such as climbing antenna towers and working with tools.

It's not expected that any particular person would encounter all these problems. But, as work assignments change frequently, your job tomorrow may place you in an area where you may have no idea of the pitfalls and related precautions.

Pointing out the hazardous equipment and operations in electronics is not meant to cause alarm but to show the dangers so that you can decide the risks you want to take.

In preparing this book, the writer has extracted pertinent safety information from safety magazines, government reports, national standards, manufacturers' literature, and interviews with engineers, technicians, and safety professionals.

Corrections and suggested changes will be welcomed. Please send to the author, Mr. Edward A. Lacy, 356 Claridge St., Satellite Beach, Fla. 32397.

E.A.L.

# HANDBOOK OF
# ELECTRONIC
# SAFETY PROCEDURES

# 1

# INTRODUCTION

In everything we do, perhaps, there is an element of risk. Obviously some occupations are more dangerous than electronics, but just how much more dangerous?

According to the Department of Labor, 2 million disabling work injuries occur every year in this country. An estimated ninety thousand of these injuries result in permanent disability. Worse still, more than one hundred thousand people die from such injuries. In the field of electronics, fortunately, the statistics are not nearly so grim. Less than fifteen hundred people die as a result of electrocution and lightning and many of these, of course, are not involved in electronics either as a hobby or as an occupation.

Does this mean then that there are no hazards, obvious or hidden, in the field of electronics? Is it safe to ignore those who constantly warn about the dangers of lasers, high intensity noise, new chemicals, shop tools, microwave and X-ray generators, tall antennas, underground cable tunnels, lightning, static electricity, and high- and low-level voltages? The answer is an emphatic "no!" First, statistics do not tell the whole story. Today's near miss may be tomorrow's fatality, but near misses are rarely reported. For example, many technicians receive nonfatal electric shocks during their careers, but few are likely to report these occurrences; both the technicians and their superiors attach little importance to accidental shocks.

Second, statistics take on a different meaning when *you* become one of the statistics. The risks are definitely there, as we will show and as you may find through ignorance and carelessness.

Why keep safety in mind while you work? Besides avoiding the pain and discomfort of an injury, you owe it to your coworkers, your family, and your wallet.

Severe accidents can wreck your personal finances. If you have an accident, Workmen's Compensation is likely to pay you only a paltry amount each week you are disabled. Although efforts are being made to increase payments, the fact remains that the majority of states pay less than $100 per week at this time. Benefits vary from a mere $49 per week to the more reasonable, but rare, amount of $167 per week.

If you disobey safety rules, you may be fired from your job under the new Occupational Safety and Health Act, which is summarized in the following paragraphs.

## OCCUPATIONAL SAFETY AND HEALTH ACT

Whether you're an employer or an employee, you should know the basic provisions of the Occupational Safety and Health Act of 1970, which went into effect on April 28, 1971. As far as the working man is concerned, this act is probably one of the most important laws to be passed in this century, for it requires employers to furnish safe and healthy working conditions, free from recognized hazards. In effect, it requires, in principle at least, just about every employer in the country to provide safe working procedures and proper safety equipment for his worker. (However, some critics of OSHA point out that it is *not* doing the job it was set up for, partly because it is seriously underfunded.)

Before the act was passed, some businesses and trade associations waged a bitter fight to keep parts of it from being enacted, primarily because they were unwilling to change their methods and to invest money in safer equipment and buildings for their workers. Indeed, some of them continue to drag their feet and contest parts of the law; fortunately, however, many employers consider it their moral obligation to abide by the law and are spending a lot of money in the attempt.

The provisions of the law apply to 57 million workers in 4 million establishments. For all practical purposes it covers all businesses, large and small, that are engaged in commerce. The government's definition of *commerce* applies to just about all of us, except for coal miners, atomic energy workers, and government employees who are covered by other laws.

Under the law, each *employer* must:

1. furnish each employee employment and places of employment that are free from recognized hazards causing, or likely to cause, death or serious physical harm.
2. comply with safety and health standards issued under this law.
3. comply with record-keeping requirements of this law.

*Employees* have the duty to comply with these same safety and health standards, rules, and regulations. If, as an employee, you do not comply with these standards, you may, by law, be dismissed or otherwise disciplined by your employer. And if you are dismissed you may not be able to receive unemployment compensation.

To make sure employers do their part, the Occupational Safety and Health Administration (OSHA), part of the U.S. Department of Labor, has set up regional offices throughout the country, staffed by hundreds of compliance officers who inspect businesses, industries, and construction sites (see Figure 1-1).

In enforcing these standards, OSHA compliance officers may enter any establishment covered by the act to inspect the premises and all pertinent conditions, structures, machines, apparatus, devices, equipment, and materials therein. Also, the compliance officer may question privately any employer, owner, operator, agent, or employee. If the officer determines that the act has been violated and issues a citation, the employer may be fined up to $10,000 for each violation. At the same time OSHA will prescribe a time for eliminating the hazard; if the employer fails to meet this date, he may be penalized up to $1,000 each day the violation persists.

In a typical month, November 1974, OSHA made 6,458 inspections and issued 4,686 citations with proposed penalties totalling $690,332. (During 1972 approximately one out of five establishments passed inspection.) As of November 1974, OSHA had made 185,000 inspections, resulting in 124,029 citations. Penalties for these citations were $16.1 million.

Because there are not enough OSHA compliance officers, and these concentrate on high risk industries, it may be some time before they get around to your shop or plant. However, they do make random checks, so if you believe you have plenty of time to correct safety hazards, you may be making an expensive mistake.

If an employee believes that a violation of a job safety or health standard exists that threatens physical harm or that an imminent danger exists, he may request an OSHA inspection. When an employee exercises his rights under the act, the act protects him from discharge or discrimination. That is, if an employee reports his employer for a serious safety violation, the employer cannot legally retaliate.

In general, OSHA's job safety and health standards consist of rules for avoidance of hazards which have been proven by research and experience to be harmful to personal safety and health. Some say that these rules are just common sense. Many of them have been around for years, but only as voluntary standards. At any rate, they are minimum, not optimum, safety requirements.

Whether OSHA's efforts will have their intended effect may not be known for years. Some critics have been disappointed because OSHA has been

Region VI
Texaco Building
1512 Commerce Street
Dallas, Texas 75201

Region VII
911 Walnut Street, Room 300
Kansas City, Missouri 64106

Region VIII
Room 15010, Federal Building
1961 Stout Street
Denver, Colorado 80202

Region IX
Box 36017
450 Golden Gate Avenue
San Francisco, California 94102

Region X
Smith Tower Building, Room 1808
506 Second Avenue,
Seattle, Washington 98104

Region I
18 Oliver Street
Boston, Mass. 02110

Region II
Room 3445, 1 Astor Plaza
New York,
New York 10036

Region III
15220 Gateway Center
3535 Market Street
Philadelphia, Pa 19104

Region IV
1375 Peachtree Street N.E.,
Suite 587
Atlanta, Georgia 30309

Region V
230 South Dearborn Street
32nd Floor
Chicago, Illinois 60604

**Figure 1-1.** Regional offices of the Occupational Safety and Health Administration.

underfunded and because workers have not taken advantage of their rights and have not asked pertinent questions on job-related safety. Others complain that OSHA overemphasizes recordkeeping and that standards are poorly written. On the other hand, OSHA has made thousands of plant managers aware of safety, some perhaps for the first time. In particular, smaller companies are now initiating safety programs. Expenditures for on-the-job health and safety were 26% higher in 1973 than in 1972, according to a McGraw-Hill Publications survey, for a total of $3.16 billion. Short courses on OSHA requirements are given at irregular intervals throughout the country by OSHA. Contact your nearest OSHA office for schedule.

## GENERAL RULES FOR SAFETY

1. Accidents don't just happen—they are caused by ignorance, fatigue, pressure, faulty or improper tools, wrong procedures, and carelessness.
2. Since people resist change, it is easier to alter equipment and tools than the habits of workers.
3. Laboratory work by its very nature is not readily adaptable to ordinary safeguards and precautions says Navy Manual NAVMAT P5100, which makes the following observations:

Much of the work will be in relatively unexplored fields in which the characteristics or behavior of the subject are yet to be determined. The operator of every test or equipment should carefully analyze the procedure to be followed and should take all necessary steps to protect himself and his fellow workers from any possible accident.

4. Although some safety hazards are obvious, others must be pointed out and explained. Workers must be convinced of the need for safety precautions.
5. As new employees may be ignorant of hazards and safety procedures, they must be told what is to be done, how to do it, shown how to do it safely, given proper tools and protective equipment, and then supervised for safety.
6. People are more likely to follow easy, convenient safety rules than cumbersome, time-consuming ones.
7. Be emotionally prepared to do whatever task is before you; if you're under stress, you're more likely to have an accident. Keep your mental pressures under control.
8. Do not distract fellow employees during hazardous work or allow them to distract you.

9. When several people are involved in a hazardous operation, group pressure or ridicule may keep some from following safety rules. One solution may be to get the group in on the decision.

10. If you have more than your share of accidents, if you're accident-prone, look for underlying physical and psychological reasons. "Accident proneness" is not an incurable disease.

11. Remember Murphy's law—if anything can go wrong, it will, sooner or later.

12. If you have an on-the-job accident, notify your supervisor. If you're self-employed, notify your insurance agent and complete the applicable OSHA forms.

*Twelve Concepts of Safety*[1]

1. Safety is the same in all industry—Where coal mines have the best record, they use the identical safety program that is effective in steel or other industry.

2. Safety is primarily a people problem—Most accidents are caused by faulty acts of people rather than unsafe conditions. The problem is to get people to do the right thing and keep them doing it.

3. Safety is no accident—Not just luck. A good, planned program will get results.

4. Safety pays—Good safety practice is not just an added cost—it is an insurance investment for profit.

5. A well-established company safety policy is important—Let the employees know that work safety is of primary concern, and that the company is continuously working to improve it.

6. The company must assume responsibility for all accidents—Not, "why did the idiot hurt himself," But, "what have we done wrong to allow this to happen." Management must assume responsibility because it controls the resources and sets the priorities.

7. Determine the cause rather than place the blame—Everyone will help find the cause. No one wants to accept the blame. If blame is the object, then the real cause becomes obscured.

8. Investigate all accidents and incidents thoroughly and properly—Determine the cause, develop a defense, and follow through. Where this procedure is used, accident rates drop—compare airline fatality rates with highway fatality rates. An investigation is a teaching process.

[1]From "OSHAct What It Should Be," by Elmer A. Fike, *National Safety News*, March, 1973.

9. Don't blame the careless worker—This excuse leads nowhere. There will always be careless workers. (Prevent accidents in spite of careless acts.) Educate the worker to a "faith" in safe practices.

10. Motivate rather than penalize—Motivation is almost always more effective. Sell the men on the program, get them on the safety bandwagon.

11. Every man a job safety inspector and expert—Don't depend on the staff safety man or outside inspectors. Every one must be watching and practicing safety on the job.

12. Knowing is not enough—The concepts of safety must be put into practice.

## GUIDE TO THIS BOOK

Whether you are an electronics technician, engineer, hobbyist, or ham operator, you should know about hazards in electronics and how to protect yourself. (Some of the hazards, it should be noted, are not at all obvious.)

This book provides pertinent safety precautions and procedures for all who repair, design, install, or use electronic equipment. Basically, the book covers two categories of hazards:

1. Those which are the result of electronic equipment being built, operated, repaired, tested, or installed (for example, shock and radiation).

2. Hazards not unique to electronics but which can reasonably be expected to be encountered by electronics personnel during electronic equipment installation and maintenance (lifting, climbing, digging, and using hand tools).

This book is personal—it's written for the individual employee or employer. It is useful only incidentally to a company safety director. There are no chapters on setting up a safety program or keeping safety records or making reports. Nor will the reader be belabored with the obvious rules of good housekeeping; he doesn't need a book to tell him that greasy or wet floors or loose objects in walkways and on stairs are dangerous.

Some of the chapters concern specific groups of workers. For example, *Underground Hazards* applies almost entirely to CATV technicians. But job duties in electronics can change rapidly; it's conceivable, for instance, that a radar technician may be called upon to install underground cable or to maintain such cable. Because the radar technician does such maintenance so infrequently he is likely to be unaware of necessary safety precautions which may be second nature to a CATV technician who is in daily contact with these hazards.

## DISCLAIMER

To the best of our knowledge, the safety precautions outlined in this book should enable you to minimize the risks you may encounter in electronics work. Although these precautions may not cover all eventualities, they should, if followed, provide you with a measure of safety on the job. We have tried to include the best, most authoritative advice available on electronic safety. However, neither the author nor the publisher can assume liability for any accidents that may occur—local conditions may vary, products may have changed since publication, information originally supplied by manufacturers may be incorrect, etc. If you encounter hazards not documented in this book, do not consider them unimportant or nonexistent. Vigilance is always necessary in electronics work, particularly with new devices and equipment whose potential for causing harm has not yet been determined.

Mention of specific brands or trade names in the following chapters does not necessarily imply endorsement by the author or publisher.

# 2

# ELECTRIC SHOCK:

# UNSAFE AT ANY VOLTAGE

## INTRODUCTION

As mentioned earlier, electric shock is one of the most common hazards encountered by people employed in the installation, test, repair, and operation of electronic equipment (see Tables 2-1 and 2-2). Because they have survived such shocks, however, these people have a tendency to downgrade the danger, to minimize the problem. Unfortunately, they forget that the same set of circumstances that kept the shock from being fatal once may combine differently on the next occasion to make the shock a killer.

Absolute statistics on nonfatal shocks are hard to obtain and may never be known—not many technicians or engineers will voluntarily report that they were so careless or ignorant as to get themselves shocked—but it has been estimated that 30,000 people receive electric shocks each year.

With misleading statistics like these, the electronics worker may be tempted to ignore shock hazards and to consider shock to be an unlikely occupational hazard. To point out the danger is not meant to discourage you or to have you consider giving up electronics as an occupation or as a hobby. Rather it should encourage you to take necessary safety precautions, which are not nearly as complicated or burdensome as hospital care or funerals.

With any electrical device or circuit, there is a potential shock hazard. For anyone working with electrical or electronic circuits, it is not enough, however, to know that electricity can hurt or kill you. It's important to understand the beast, to know how and when it can shock you, if you are to protect yourself from such hazards.

**TABLE 2-1**   *Deaths caused by electric current in the United States*

| Year | Home | Farm | Industry | Street | Unspecified & Others | Total |
|------|------|------|----------|--------|----------------------|-------|
| 1973 | 303 | 72 | 174 | 92 | 508 | 1149 |
| 1972 | 292 | 64 | 138 | 110 | 484 | 1088 |
| 1971 | 288 | 59 | 146 | 83 | 489 | 1065 |
| 1970 | 277 | 83 | 168 | 124 | 488 | 1140 |
| 1969 | 318 | 71 | 146 | 134 | 479 | 1148 |
| Average | 296 | 70 | 154 | 109 | 490 | 1118 |

(Courtesy of Public Health Service, U.S. Dept. of Health, Education, and Welfare)

**TABLE 2-2**   *Shock deaths,[1] a representative sample*
(For the years 1965 through 1971)

| | |
|---|---|
| Antennas—radio, ham, CB, TV—erecting and dismantling | 125 |
| Portable drills | 85 |
| Musical instruments (electrical) | 9 |
| Public address systems | 7 |
| Radio—home | 10 |
| Television | 9 |
| Radio transmitters | 3 |
| Miscellaneous electronic devices | 9 |

[1]Francis M. McKinney, *Statistical Summaries 1965 to 1971*, The Electricide Series, 1972. (Adapted by permission of the author.)

It is true that "high voltage kills." Undoubtedly you have seen signs near electrical installations which say, "Danger: High Voltage, Keep Out" (Figure 2-1). The signs certainly mean what they say, but unfortunately they have been misconstrued by some people who have concluded that if *high* voltage kills, *low* voltage cannot or does not kill.

Actually, the amount of voltage you touch is only one of the factors involved in the degree of shock you receive. Men have been killed when they touched high-voltage conductors—10,000 volts or so. But just as many people have been killed by so-called low voltages—ordinary household voltages of 110 to 120 volts. Fatalities have occurred at voltages as low as 24 volts ac, which shows that almost no voltage can be considered safe if adverse circumstances are present (if, for example, you are sweating profusely or if you're standing in water).

Just how badly you will be affected by an electric shock depends on the following primary factors:

1. The amount of current, measured in milliamperes (*not* amperes), that flows through your body.
2. The path that the current takes from entry to exit from your body.
3. The time, measured in milliseconds, you are in the circuit.

**Figure 2-1.** Danger signs in these locations mean what they say. Do not misread—there is no implication that low voltage is not dangerous. (Courtesy of Argonne National Laboratory.)

Secondary factors include your age, size, and physical condition, and the frequency of the current—ac is slightly more dangerous than dc at the same level.

Obviously, the amount of current that flows through your body will be determined by the voltage and the resistance of your body from the point the current enters your body to the point where it leaves. In any case, the current will follow the path of least resistance. This resistance may vary from 300 ohms to 100,000 ohms, depending on the thickness of your callouses, if any, the amount of perspiration, your age, the area of electrical contact, and the path the current takes through your body.

If the skin is broken or cut, the resistance may be as low as 300 ohms. Although resistance varies from moment to moment and from person to person, 500 ohms is generally considered to be the maximum resistance that you can count on. At 500 ohms, you cannot safely touch more than 9 volts!

For shock to occur you must be part of a closed circuit in which current can flow. This is most likely to happen if you touch a hot wire while grounded yourself. It can also happen if you touch both wires of an energized circuit. With a high resistance of 100,000 ohms, about 1 mA will flow through your body if you complete a circuit across 120 volts. Most people can feel such a current and will voluntarily let go.

As the resistance decreases, the current increases. Between 1 and 6 mA you may receive just a "tingle" or perhaps an uncomfortable or painful shock, but not a fatal one. Any death from such a shock would be indirect; for example, by being startled you might fall off a ladder. At currents above 6 mA, the situation becomes deadly serious. The pain becomes intense and you may not be able to turn loose of the circuit. In effect you are frozen to it.

Experiments performed by Charles F. Dalziel at the University of California at Berkeley and W. R. Lee at the University of Manchester show the average "let-go" ac currents to be 16 mA for men and 10.5 mA for women (see Figure 2-2). As a result of their experiments, the *safe* let-go currents are now considered to be 9 mA for men and 6 mA for women.

If you are frozen to a circuit, your resistance may start decreasing and it's possible, if you are in the circuit long enough, for the current to rise to a fatal level. At 25 mA your muscles can go into violent contractions. If you're lucky, the contractions will be so great that you will be thrown clear of the circuit.

When the current through your body is between 50 and 200 mA, your heart can be thrown out of rhythm (ventricular fibrillation occurs, as shown in Figure 2-3) or stopped altogether, or your breathing can stop before your heart is affected. In some cases you could be revived by closed-chest cardiac massage and mouth-to-mouth resuscitation, provided fibrillation does not occur and provided that resuscitation is given promptly by a person sufficiently trained in such techniques. When such procedures fail, death occurs within a few minutes.

Such currents may not be fatal, however, if the duration of current flow is kept short enough. This principle is used in ground fault interrupters, which typically allow the current to flow for a maximum of 25 milliseconds. In this time period, the current does not have enough time to electrocute you.

The effect or severity of the shock depends also on the current's path through your body. If it flows from one finger to another finger on the same

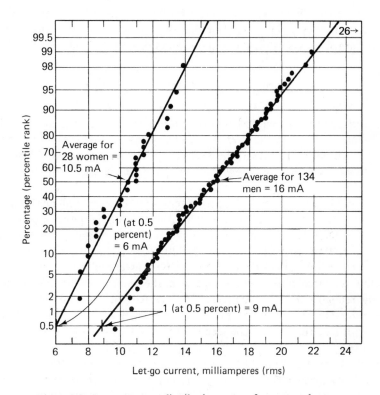

**Figure 2-2.** Let-go current distribution curves for men and women (at 60-Hz commercial alternating current). (Courtesy of © *IEEE*. From Charles F. Dalziel and W. R. Lee, "Lethal Electric Currents," *IEEE Spectrum*, Feb., 1969, p. 44.)

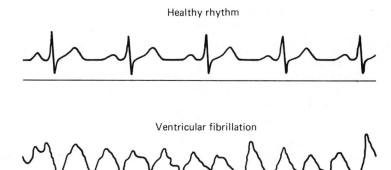

**Figure 2-3.** Heart patterns.

hand, it is much less severe than if it flows from one hand through your heart and lungs to the other hand. For this reason, technicians working on hot circuits are always cautioned to put one hand behind the back or in the pocket.

## RECOGNIZING SHOCK HAZARDS

Electrically unsafe equipment may continue to operate effectively without giving any warning to the user. However, in many situations electric shock hazards are quite obvious to the average technician or engineer. You don't have to be an expert to know that any kind of electrical equipment near swimming pools, marinas, bathrooms, basements, damp floors, or outdoors can be dangerous. It should be apparent too that any antenna you erect may fall, either during erection or later during use; if a power line is in the path of the fall, current may flow down the mast or guy wires to anyone unlucky enough to be holding on to either. It is likely that some of the victims of antenna shocks in recent years were aware of the potential danger. Possibly they misjudged the distance from the antenna to the power line, or, having recognized all the dangers they erected the antenna from the safe side of the line, only to have it fall in "the wrong direction," hitting the power line. Another situation may be seen in Figure 2-4.

Murphy's law—"if something *can* go wrong, it *will* go wrong"—should be kept in mind in any electrical or electronic installation, operation, test, or maintenance procedure. It is equally important never to assume anything about any electric circuit except perhaps that it can kill you if you don't take the proper precautions.

A few of the more obvious shock hazards are listed in Table 2-3. Table 2-4 lists some shock hazards that are not so obvious—in fact they may be

**TABLE 2-3**  *Obvious electric shock hazards*

---

Testing a circuit with a wet finger to see if it's "live"

Working on circuits assumed to be "dead" before checking that they *are* dead, and failing to prevent such circuits from being energized by others

Any electrical device that gives you a "tingle" when you touch it

Electrical equipment that does not have the Underwriters Laboratories (UL) label on both the cord and the equipment

Spliced, broken, frayed, and cracked cords and plugs

Ungrounded tools (drills, etc.), unless they are protected by double insulation

Transformer-less sets that may have hot chassis

Electric cords in floors where they may be worn or broken by people walking over them

Overloaded receptacles and circuits

Using metal ladders where they may brush or fall against power lines

Drilling into wall or floor without knowing what's on other side—it could be a hot electrical circuit

---

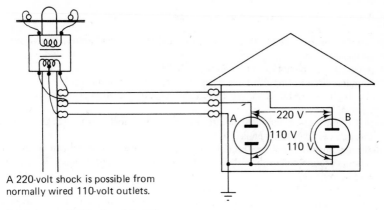

A 220-volt shock is possible from
normally wired 110-volt outlets.

Breakdown of 0.01 μF unit in (A)
or isolating capacitors in (B) may
render cabinet or exposed parts "hot".

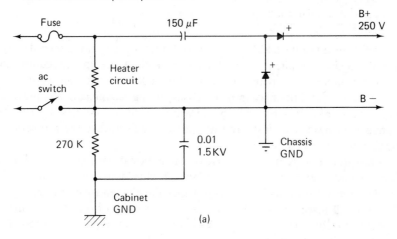

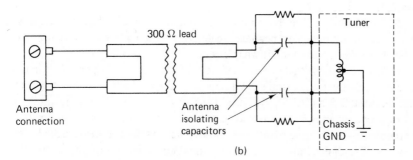

**Figure 2-4.** Possible shock hazards. (Courtesy of *Radio & TV News*.
From Kenneth Braham, "Hot Chassis? Play It Safe," *Radio & TV
News*, Dec. 1958, Copyright 1958 by Ziff-Davis Publishing Co.)

TABLE 2-4   *Subtle electric shock hazards*

Equipment bearing the UL label may become dangerous later because of environment or misuse

Equipment may be hot even though turned OFF

Unfamiliar circuits, unpredictable types of component breakdown

Hot cabinets—a result of failure to replace insulating spacers between chassis and cabinet

Hot TV antenna—component failure giving rise to shock path through set balun through twinlead to antenna

Replacing push-on knobs with knobs having metal setscrews which can become hot

Insulation around collapsible rabbit-ear TV antennas may fail; if chassis is hot, rabbit ears become hot

Items on top of TV chassis may be hot

Exposed metal screws and control shafts on equipment with plastic cases may be hot

Fallen power line that does not crackle or pop can be harmless one minute and lethal the next

Interlocks and bleeder resistors may fail

Capacitors that can discharge more than 50 joules

downright sneaky. Like a hard-to-find roof leak, they may cause their damage some distance from their source. A component failure in one equipment rack, for instance, may create a hazard in the rack next to it. New equipment with unfamiliar circuits may trap the unsuspecting technician. The previous owner or user of a device may forget to warn you of the dangers. Instruction manuals may have been lost. And of course there's always the problem of: "Familiarity breeds contempt."

In research laboratories particularly, the nationally recognized and local electrical codes and standards may not provide satisfactory safety. The Atomic Energy Commission points out that the element of the unknown is inherent in all research, and therein lies the possibility of direct and indirect electrical hazards for which there can be no previous record or experience. Thus, there must be special efforts in such labs to recognize and control shock hazards.

More recently it has become apparent that normally safe electrical and electronic equipment can become treacherous when placed in environments other than those for which it was designed.

Consider the fantastic array of new electronic equipment acquired in recent years by the nation's hospitals for use in diagnosis and treatment (see Figures 2-5 and 2-6). At some time during his stay at a hospital, the average patient may be connected to one or more of these medical electronic instruments—even several at one time if he has had a heart attack. The benefits of such equipment are unquestionable; patients and physicians alike profit from their use. Unfortunately, in some situations this equipment may deal the patient a fatal shock. Although the specific number of patients electrocuted each year in our hospitals is debatable, there is no denying that

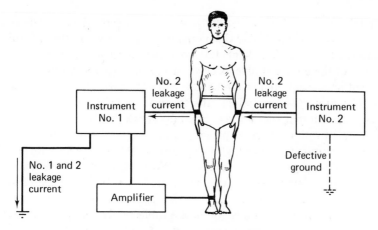

Even with feedback amplifier,
patient is in danger of shock
when more than one instrument
is used.

**Figure 2-5.** Potential shock hazard in the hospital. (Courtesy of *Popular Electronics*. From Hector French, "Medical Electronic Equipment and Hospital Safety," *Popular Electronics*, Jan. 1972, Copyright 1972 by Ziff-Davis Publishing Co.)

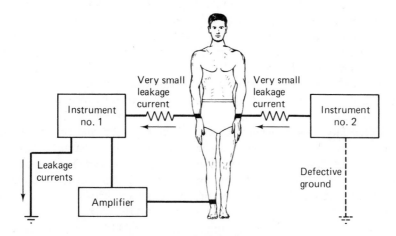

Safest hookup is shown here. In
addition to amplifier, current-limiting
high-isolation inputs are used.

**Figure 2-6.** Safe hookup for the hospital patient. (Courtesy of *Popular Electronics*. From Hector French, "Medical Electronic Equipment and Hospital Safety," *Popular Electronics*, Jan. 1972, Copyright 1972 by Ziff-Davis Publishing Co.)

hospital electronic gear can be dangerous. Instruments that are safe enough by themselves become deadly when operated together with other instruments connected to the same patient. But even one instrument can create trouble because of the unique problems encountered in hospitals. Corrosive liquids, frequent movement, and rough handling of equipment may cause insulation to fail or protective ground circuits to open. In either case the result is a leakage current that may severely or fatally shock the patient. In an ordinary environment with the average person, this leakage current might not even be felt; electrocution would be highly improbable. However, a hospital patient is likely to be in poor physical condition. Worse yet, some of the instrument leads may penetrate the skin or even the heart. Internal shock ("microshocks") can be fatal at currents as low as 10 to 20 microamperes if the current flows through the heart. Solutions to these problems are complicated and are best left to medical electronics specialists.

## TESTING FOR ELECTRIC SHOCK HAZARDS

An excellent way to prevent electric shock is to have equipment inspection and preventive maintenance on a regular basis and periodically test the equipment for hazards.

Regular inspection can ensure that all equipment and extension cords are free from abrasion (frayed or cut insulation), splices, moisture, and oil. When cord insulation begins to crack or flake, replace the cord. If the cord is in a position where it can be stepped on or otherwise abused, move it. If it overheats, use larger wires.

Through preventive maintenance you can keep equipment free from the dust, dirt, and oil that may cause it to overheat, thereby breaking down the insulation.

Continuity tests for ground circuits, insulation tests, and current leakage tests are essential in locating and confirming that shock hazards exist. For the continuity tests, check for zero ohms resistance between the ground prong of the equipment plug and the frame and metallic parts of the equipment. For the insulation test, check for proper resistance between the frame or ground prong and the hot and neutral prongs on the plug.

To confirm that a proper ground circuit is available at the power receptacle, use an outlet circuit tester such as the Hubbell 5200 (see Figure 2-7A and B). If the tester shows a fault in the power supply receptacle, it is evidence of an obvious shock hazard which may turn an otherwise safe device into a killer.

To measure leakage current, use a Simpson Model 229 ac Current Leakage Tester (see Figures 2-8 and 2-9A) or equivalent tester (see Figure 2-9B and 2-9C).

How it operates

The Hubbell 5200 Outlet Circuit Tester has a simple arrangement of neon lights which visually indicate and identify various fault conditions in electrical circuits. (See chart below). Operation is quick and easy. By plugging the 5200 Tester into a single phase, 125 V., 2-pole, 3-wire outlet; such as those shown at left, the combination of lighted and/or unlighted lamps will immediately indicate circuit condition.

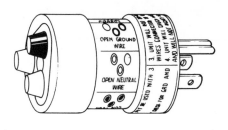

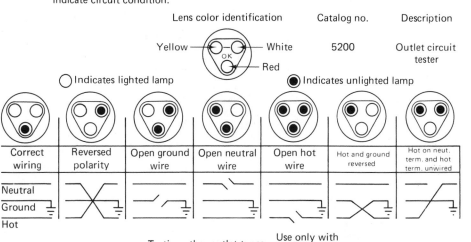

| Lens color identification | | Catalog no. | Description |
|---|---|---|---|
| Yellow —⊙ White / Red | | 5200 | Outlet circuit tester |

◯ Indicates lighted lamp          ⬤ Indicates unlighted lamp

| Correct wiring | Reversed polarity | Open ground wire | Open neutral wire | Open hot wire | Hot and ground reversed | Hot on neut. term. and hot term. unwired |
|---|---|---|---|---|---|---|

Neutral
Ground
Hot

Testing other outlet types          Use only with polarized adapters

The Hubbell 5200 Outlet Circuit Tester may be used to check many other two and three-wire outlet types wired for single phase, 125 Volt service by the use of proper adapters.

Polarized by means of one wide blade

No. 6273L          Outlet types

(a)

(b)

**Figure 2-7.** (a) Hubbell outlet circuit tester. (Courtesy of Hubbell Wiring Device Division.) (b) Completest AC outlet tester. (Courtesy of ECOS Electronics Corporation.)

19

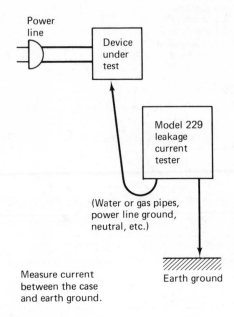

**Figure 2-8.** Test equipment setup for current leakage test. (Courtesy of Simpson Electric Company.)

Leakage current, as defined by Underwriters Laboratories Standards,[1] is "all currents, including capacitively coupled currents, which may be conveyed between accessible parts of a cord-connected appliance and ground or other accessible parts of the appliance." (See Figure 2-10.)

Whether a leakage current gives you a "tingle" or a painful jolt or no sensation at all depends on how well grounded you are and the insulation condition of the appliance. (When touching appliances, it is not advisable to be well grounded.)

The permissible leakage current for 120-volt appliances with 2-wire portable cord connection was set at 0.5 mA in 1971. To test for this leakage, perform the following steps:

1. Test with appliance ON and then OFF.
2. For 2-wire appliances, make above test; then reverse plug in outlet and repeat the test.
3. Test heating and cooking appliances when ON but cold; then test at maximum heat.
4. Test motor-operated appliances at no load; if there are speed controls, test at low, medium, and high speed positions.

[1]Underwriters Laboratories, Inc., Standard for Electric Heating Appliances, UL 499–1968.

(a)

(b)

(c)

**Figure 2-9.** (a) Simpson Electric Company Model 229 leakage current tester. (Courtesy of Simpson Electric Company.) (b) Systems research shock hazard and leakage current detector model 40. (Courtesy of Systems Research, Inc.) (c) Outlet and leakage analyzer. (Courtesy of ECOS Electronics Corp.)

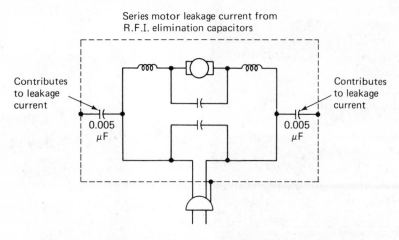

**Figure 2-10.** Potential contributor to leakage current. (From John Stevenson, *Leakage Current from Appliances*, *IEEE* Conference Paper 71CP342IGA, May 4, 1971, Underwriters Laboratories, Inc.)

5. Test regular appliances within five seconds after energizing and again after reaching thermal equilibrium.

## SHOCK PRECAUTIONS

Many methods are available for avoiding electric shock, as will be indicated in the following paragraphs. As a minimum for electrical safety, it is wise to follow the provisions of the National Electrical Code and use electrical equipment that carries the Underwriters Laboratories (UL) label. Above all, good safety precautions require an awareness of potential dangers and an unwillingness to compromise on safety.

### Insulation

Insulation is one of the oldest and best methods of protection against shock. This protection can be provided by insulation of the conductor, insulation of the worker, or isolation of the conductor from the worker. Any of these methods, however, can fail.

If the electrical circuit is physically isolated, for example, several inches or feet of air provide an extremely high resistance (that is, good insulation) unless the separation is accidentally bridged by an antenna mast, guy wire, or metal ladder.

In conductor insulation, the environment may cause the insulation to flake off or become porous. Insulation that survives the environment may fail from abuse. Cords that are disconnected by a yank on the cord instead of the plug, cords that are stepped on, run over, or caught in swinging doors, and cords chewed by rodents are likely to crack or break and provide zero protection.

As another example, a perspiring technician standing barefoot on a wet or even a damp floor has no appreciable insulation and provides an easy path for ground fault currents.

Procedures for guarding against these insulation failures are simple and straightforward.

Insulation resistance on motors should be checked every six months with a megohmmeter (such as Megger ®) and recorded. Gradual deterioration or loss in insulation will then become apparent and the motor can be repaired or replaced before it becomes a shock hazard. Insulation resistance should also be checked any time equipment is repaired.

Wherever possible, use double-insulated tools. (See Figure 2-11.) Double insulation consists of two layers of insulation—the usual functional insulation found in tools plus an independent insulation system. Together they

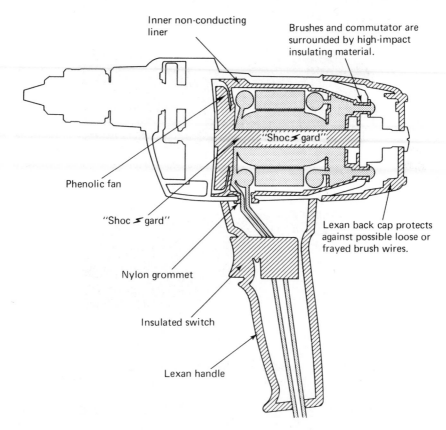

Inner non-conducting liner

Brushes and commutator are surrounded by high-impact insulating material.

"Shoc⚡gard"

Phenolic fan

"Shoc⚡gard"

Lexan back cap protects against possible loose or frayed brush wires.

Nylon grommet

Insulated switch

Lexan handle

**Figure 2-11.** Millers Fall double insulation electric drill. (Courtesy of Millers Fall.)

give an impressive safety record—it's not likely that both layers will fail at the same time. Imperfections can occur, however; if there is a defect in the cord or plug, you may still get a shock even if the tool has double insulation. The same holds true if the tool gets wet.

Tools protected by an approved system of double insulation require only 2-prong receptacles, not the 3-prong type required for grounded tools (double insulated tools are approved for residential and industrial use). Don't assume that a tool has double insulation—look for a positive indication that it is so insulated.

To provide isolation protection, use a wooden, or better still a fiberglass, ladder, so that if you tangle with an overhead power line, you have a chance of surviving. Aluminum ladders are lighter than wooden ones but unfortunately they conduct electricity. Fiberglass step and extension ladders are lighter than wooden ladders and are not affected by weather. They cost more initially but will last longer than wooden ladders.

Rubber gloves (Figure 2-12A) designed for electrical work provide additional protection around live conductors.[2] *Any tears or abrasions will obvi-*

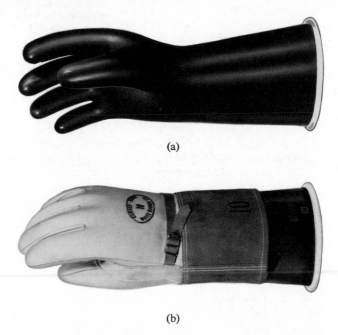

(a)

(b)

**Figure 2-12.** (a) Curved rubber hand glove, designed for natural hand shape for safety and comfort. (Courtesy of White Rubber Co.) (b) Curved-hand leather protector glove—designed to fit over the curved-hand rubber glove and for protection from cuts, tears, and punctures to the rubber.

[2]Rubber protective equipment for electrical workers must meet ANSI standards.

*ously make them valueless.* Therefore the gloves should be tested periodically by an electrical testing laboratory or possibly by the local power company. Leather gloves (Figure 2-12B) should be worn over them to protect the rubber.

Insulated floor mats should be standard equipment near any workbench or area where repairs are being made on energized equipment. The mats should be dry, without holes, and contain no conducting materials such as nails or screws. On voltages below 600 volts, the Navy recommends either dry wood platforms or nonconducting rubber mats. (Wood that has been obtained directly from a lumber yard or that has been stored outdoors is not likely to be sufficiently dry.) Whatever the material, the mat must be kept dry and clean and free from dust, solder, metal chips, etc.

The General Electric Company recommends the following insulation checks during television repair:

*Before installing the cabinet back*
1. Be sure that no lead touches a power resistor (two watts or more).
2. High voltage connections must have no sharp points.
3. The insulation on antenna leads should not be damaged. The leads should not be dressed close to any high voltage point or ac line connection.
4. The ac wiring should be inspected for damaged insulation, frayed wires, pinched leads, or cold solder connections.
5. Inspect the ac line cord for broken or damaged insulation.
6. Check for dc continuity (zero ohms) from the large pin on the power interlock to the chassis with the ON-OFF switch in the OFF position.

*After installing the cabinet back*
1. Connect the VHF antenna to the VHF antenna terminal.
2. Do not plug the receiver into a power outlet. Connect both blades of the power plug together and place the ON-OFF switch in the ON position.
3. Measure between the shorted power plug and the following points. Readings should be as indicated (typical values are given; these can vary from set to set):

| | |
|---|---|
| Antenna terminals—UHF | 600 K min, 4.4 megs max. |
| Antenna terminals—VHF | 600 K min, 4.4 megs max. |
| Cabinet back screws | Open circuit |
| Channel selector knobs | Open circuit |
| All secondary knobs | Open circuit |
| Metal escutcheons and overlays | Open circuit |

If any reading is outside limits, fault must be corrected.

### *Grounding*

In electrical equipment or systems, a *ground* is an electrical path to earth. A deliberate ground will have a low resistance. If it is unintentional it may have a much higher resistance.

In an intentional ground the path to earth is a large diameter copper wire connected to an all-metal underground cold-water system or a special 8-ft-long rod driven into the earth.

In the unintentional ground the path to earth may be through wet concrete floors, radiators, kitchen cabinets, etc. Compared to ground rods or water pipes, these grounds are inefficient, but they are still effective enough to electrocute the hapless person who touches an electrically "hot" device at the same time he touches an unintentional ground.

In most homes the neutral wire from the power line transformer is grounded at the service entrance to the house. This *service* or *system ground* protects the house wiring from power surges from lightning, etc. By itself, however, it is useless when appliances break down and become electrically hot.

An *equipment ground* is a separate conductor, a third wire, connected to the appliance or tool frame or metallic parts. It runs through the house wiring back to the system ground. Theoretically it provides a low resistance path from the frame of the appliance to earth, and thereby diverts any stray currents *directly* to ground. By providing an easy path to ground for this current, the equipment ground effectively prevents the current from going through any person using or touching the device.

Under ideal conditions, of course, there would be no need to ground the shell of an appliance. But insulation does crack and break, sometimes from old age, sometimes from overheating. Moisture, dirt, and carbon particles may then get in and form an electrical path from the break in the insulation to the frame. Unless this fault current is shunted off to ground, the frame will have an electric potential that can be dangerous or fatal to anyone who touches it. In a situation where the equipment is grounded, the fault current is high, and the resistance of the equipment ground is low enough, the circuit breaker or fuse for that circuit will blow. Whether the fuse blows or not, the equipment ground will protect the user.

Because of the excellent protection they afford (at least when the ground connections are properly made), equipment grounds should be provided for the following equipment (unless they are specifically labelled "double insulation"): portable electric tools (including outdoor equipment), electric signs, chassis and frames of communications receivers and transmitters, home TV antenna towers, and electric equipment in garages. This list is necessarily incomplete; electric codes and obvious safety practices may suggest other items that need equipment grounds.

However, equipment grounds should *not* be provided for devices that have double insulation or open heating coils (toasters, broilers, and radiant heaters). To ground the exterior of these heaters would increase, rather than decrease, the shock hazard.

~ *Unintentional grounds such as metal chairs, workbenches, and ladders should not be used near electrical circuits.*

Electrical equipment that is apparently dead may retain a deadly charge, sometimes for hours after the equipment has been shut down. Because of this situation, all technicians should learn how to use another type of ground, called a *grounding* or *shorting rod* (or *bar, stick, prod*—see Figure 2-13). The shorting rod is a simple home-made device of wire, clamps, and a dry stick or board (better still, a bakelite rod) that should be on every work-

**Figure 2-13.** Use of grounding (shorting) stick (hook). (Courtesy of Lawrence Berkeley Laboratory, University of California, Berkeley.)

bench. The part held in the hand must be well insulated to keep you from being shocked. The clamp must make a good gripping contact to the nearest sure ground; it must not come loose easily as it might break the ground connection just at the instant you need it. (For this reason, a spring-type clip should not be used for the ground end.) The "business end" of the stick should be a short piece of wire or clamp (such as alligator clips). If it is wire, it should be formed into a hook. The wire from the ground clamp to the shorting stick must be a large conductor.

In use, connect the ground clamp to earth first. While holding the shorting stick on the insulated portion, touch the business end to the point to be discharged and leave it hooked or clamped to that point until your work is done.

Capacitors, chokes, and transformers with high capacity or inductance are particularly dangerous. When the equipment is turned off, and preferably unplugged, it is advisable to short (discharge) the items mentioned plus any other obvious item that may retain a shock potential. If it is not possible to leave the shorting stick in place until repairs are complete, discharge the capacitors three separate times, 15 seconds apart, inasmuch as the initial discharge will not get rid of all the charge. Unless this is done, you may receive a fatal shock from the charge stored in the capacitors, even though the equipment is OFF and unplugged. Bleeder resistors do not always work.

When equipment grounds are used, all parts of the grounding system must be working properly if safety is to be ensured.

An underground cold-water pipe system makes an ideal ground and is preferred, provided that it is a continuous metal system. If there are sections of plastic pipe (couplings, for instance) in the system, it will no longer make a good ground. Even with a metal system, a wire jumper must be used to bypass the water meter so that the house's ground ties into the complete city's water system.

Where such a ground system is not available, the National Electrical Code permits the use of driven ground rods. This rod must be 8 feet long, preferably copper, and at least 5/8 inch diameter. Because of its high resistance compared to the water pipe system (the earth around the electrode may be dry), this method is not highly recommended. Professor Charles F. Dalziel notes: "Almost without exception, a driven ground rod will not permit a current flow sufficient to operate the circuit protective device" (*IAEI News*, July 1969). (By the same token, a household TV tower or mast will not be adequately grounded just because its base protrudes into the earth.)

Whichever type of ground is used, it is connected to the neutral wire and the house ground wire.

The National Electrical Code specifies the proper sizes for grounding wires. The wire should be continuous, have as low a resistance as possible, and be capable of carrying the overload current in case of a fault. If the current-carrying capacity of the ground wire is too low, the ground wire may burn out without activating the fuse or circuit breaker.

Older homes may not be wired with these three-prong (hot, neutral, and ground) conductors. Generally, if a house has three prong receptacles, it has a ground at the receptacle. Some houses have three-wire conductor systems but two-prong receptacles. This situation can be determined by removing the wall plate to the receptacle; if there is a bare copper wire hanging from the power cable, then it is a simple matter to convert from two-prong receptacles to three-prong. If there is no ground wire in the outlet box, it may be necessary to run new wires back to the circuit breaker or fuse box.

Getting a good ground, maintaining it, and making sure that it is used properly can be a very difficult job.

Professor Charles F. Dalziel (*IAEI News*, July 1969) notes that defects in the grounding system are the most common cause of injury in low voltage shock accidents. He says: "Although a noble experiment, accident experience with portables supplied by three wire cords has been a great disappointment."

The most common of these defects is the habit of some workers to remove the third prong—the tool ground—when they are unable to find a three-prong receptacle. (In an emergency a three-hole to two-hole adapter may be used if you know that the screw on the faceplate of the receptacle is grounded and if you attach the pigtail of the adapter firmly to this screw. Such an arrangement is better than no ground, but not much better. It should not be used on a permanent basis. See Figure 2-14.)

Even if three-prong receptacles are available, the worker has no quick visual way of knowing if the ground system is working properly. The ground wire could be broken somewhere in the system; the break could be in a molded plug or in the house wiring and therefore would not be visible.

It's possible, too, for the grounding system to corrode and develop an unsafe level of resistance. The resistance of the grounding wire plus the water pipe connection should not exceed 3 ohms. (According to some experts, 1 ohm is the resistance allowable to avoid shock; others feel that 0.1 ohm is the safety level.)

For this reason, tools and receptacles should be tested at least twice a year.[3] The common ohmmeter is not satisfactory for ground resistance testing because of stray ac and dc currents. Instead, use a Biddle Co. Megger[4] ® Null Balance Earth Tester (Figure 2-15).

The tool or appliance can be checked with an ordinary ohmmeter. There should be zero resistance between the ground prong and any exterior metal

---

[3] Use an outlet tension tester to ensure proper tension on appliance plug blades.

[4] Megger ® is a registered trademark of the James G. Biddle Company.

parts. There should be at least 100,000 ohms resistance between the ground prong and either of the other two prongs on the power plug.

If you replace a plug or cord on a grounded appliance or tool, it is absolutely essential that you do not reverse the wires. The green wire must be connected to the ground prong on the power plug and to the frame or shell of the appliance (see Figure 2-16).

When using grounded tools or appliances, recognize that they create a possible danger exactly at the time they are grounded efficiently and prop-

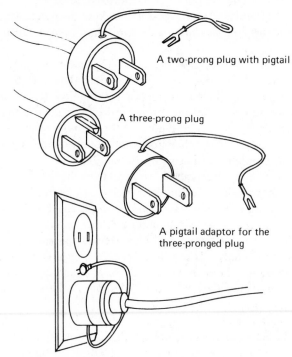

Use ground wires

Electricity follows the path of least resistance to ground. Faulty wiring in an applicance can energize an appliance, and you can become the path of least resistance. Always ground appliances. There are three ways:

A two-prong plug with pigtail

A three-prong plug

A pigtail adaptor for the three-pronged plug

You must connect the pigtail. And you must have grounded outlet boxes for any of the three methods to protect you.

(a)

**Figure 2-14.** (a) Proper use of ground wire adapter. (Courtesy of the National Safety Council.) (b) Use the correct plug.

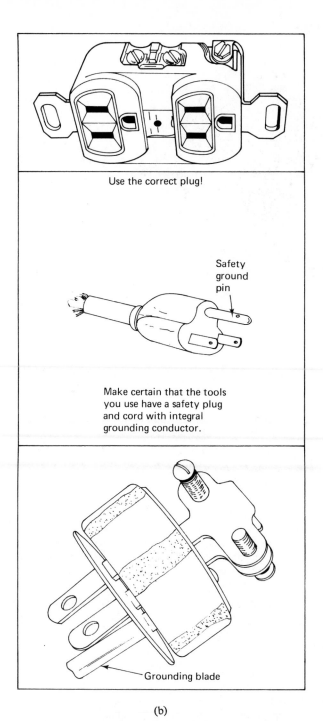

Use the correct plug!

Safety ground pin

Make certain that the tools you use have a safety plug and cord with integral grounding conductor.

Grounding blade

(b)

**Figure 2.14.** (Continued)

**Figure 2-15.** Null-balance earth tester (or Megger ®). (Courtesy of James G. Biddle Company.)

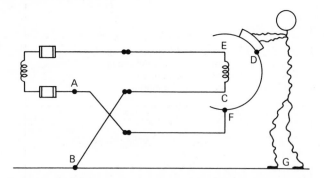

**Figure 2-16.** Wrong connections in receptacle, perfect insulation on tool and system; result: shock. (From Lt. R. L. Kline and Dr. J. B. Friauf, *Electric Shock—Its Causes and its Prevention*, U.S. Department of the Navy.)

erly. The danger is that you may have many more opportunities to touch ground if you are using an appliance or tool that is not properly grounded. *Remember: If it isn't grounded, it isn't dead.*

### Ground Fault Interrupters

Whenever insulation, double insulation, isolation, and grounding fail to do their job of keeping electricity in its place, an extremely dangerous situation may develop wherein the exposed parts (frame or shell) of an electrical device develop an unintended electrical potential. By touching such a device, you will establish an unintended path for electric current between the ungrounded conductor and ground. In this ground fault condition, electrocution is very likely to occur, unless your body resistance is high. Regardless of your resistance, no great harm will be done if the circuit is protected by a *ground fault interrupter* or *ground fault circuit interrupter* (Figure 2-17).

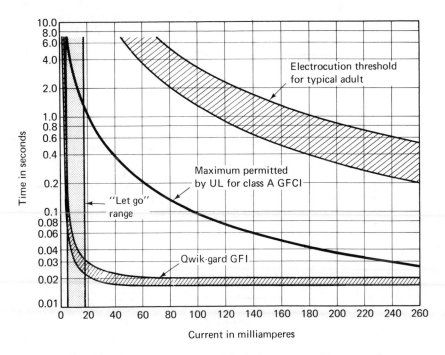

**Figure 2-17.** Action of ground fault interrupter. (Courtesy of Square D Company.)

The ground fault interrupter (hereafter called *GFI*) is a supersensitive, rapid-action power switch that does nothing until it detects the ominous fact that current is leaking to ground; whereupon, in a few milliseconds, it turns off the whole circuit (see Figures 2-18 and 2-19). With a GFI in the circuit, you may receive an uncomfortable shock from a ground fault, but the shock will not kill you, simply because you are not in the circuit long enough.

For the twenty years that GFIs have been in use in other countries, the National Safety Council reports that there has not been a single recorded electrocution from ground fault in any occupancy employing this type of protection. In 1971 A. W. Smoot of the Underwriters' Laboratories estimated that 81 % of the electrocutions in the preceding forty-five months might not have happened if the circuits involved had been protected by GFIs.

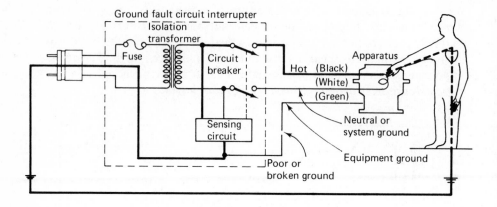

An isolation-type ground fault circuit interrupter provides for the fault itself to trigger an electronic switch to open the circuit breaker. The isolation transformer, as the name implies, isolates the employee-user and provides back-up protection against electronic component failure.

**Figure 2-18.** Isolation-type ground fault interrupter. (Courtesy of National Safety Council. From Allan Reed, "Ground Fault Circuit Interrupters," *National Safety News*, Dec. 1969.)

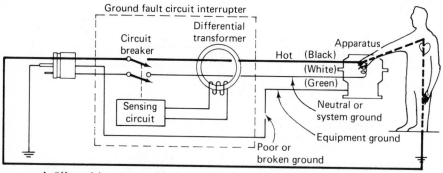

A differential-type ground fault circuit interruptor reacts to an imbalance created when tiny amounts of current seek ground without passing through the doughnut-shaped transformer. The imbalance is amplified by the sensing circuit that opens the circuit breaker.

**Figure 2-19.** Differential-type ground fault interrupter. (Courtesy of National Safety Council.)

Official recognition and validation has been given to GFIs by the National Fire Protection Association: the National Electrical Code specifies that all 120-volt, single-phase 15 and 20-ampere outlets installed outdoors for home use must have GFI protection.

Because of the GFI's excellent features, it is tempting to consider the GFI a panacea for all electrical hazards and to ignore or dispense with routine safety procedures. To do so, however, would be more than foolish; it could be disastrous. The GFI does not take the place of proper grounding. (For example, if you're drilling into a wall and hit a hot conductor with your drill bit, a GFI in your drill circuit will not protect you.)

Reliable though they are, GFIs can fail. For this reason they all have self-test provisions, and it is recommended that you check them monthly, or even before each use on some models.

Finally, it is most important to keep in mind that the GFI only provides protection from electric shock resulting from *line-to-ground* contact. If you are part of a *line-to-line* contact, if you are touching both the hot wire and the neutral wire, there is no way the GFI can protect you. Fortunately for possible shock victims on GFI circuits, it's estimated that up to 90% of electric shocks occur from line-to-ground contacts, the very situation GFIs were designed to protect.

Although GFIs are made in various shapes and sizes, there are only two basic types: isolation transformer and differential transformer.

Woodhead's Portable Ground-Fault Sentry uses an isolation transformer as backup protection for its ground fault sensing circuits and also as a means of avoiding nuisance tripping (from power surges, etc.) at its very low tripping level of 0.2 mA.

In the differential transformer models, the most common type, power lines to the load are run through a doughnut-shaped differential transformer. In a normal situation, the current in each wire is the same and the GFI in effect coasts along. When a ground fault occurs, some of the return current from the load bypasses the GFI on its way to the electric service entrance ground. (Note that this leakage or runaway current can be large enough to electrocute you but too small to blow a fuse.) In an amazing 25 milliseconds or less, the GFI senses that the current in the hot line is not the same as in the neutral line and proceeds to shut down the entire circuit. This fast response is quick enough to keep you from being electrocuted, although you may feel the shock.

Except for Woodhead's Portable Ground-Fault Sentry, most personnel-type GFIs are set to trip at 5 mA or less, the level prescribed by Underwriters Laboratories. As long as the ground fault current stays below 5 mA, the GFI will not operate, and you may still receive a nuisance shock. (For this reason people with heart problems cannot rely on the standard GFIs to protect them.) At a trip level of 5 mA most people can still voluntarily let go of a hot wire or appliance. The 5 mA-level is a compromise between providing maximum protection and avoiding undesired tripping. Should the level be set much lower, in many instances the GFI would continually turn off the circuits because of cumulated leakage from several tools or appliances and also because appliances built before 1969 were allowed to have leakage currents up to 5 mA.

GFIs are available in portable, semiportable, and permanent-installation models. The permanent-installation models include various wall-mounted box types such as those made by Hubbell (as in Figure 2-20), a miniature version combined with a conventional duplex receptacle, and a miniature version combined with a circuit breaker [Square D Company's Qwik-Gard (TM) see Figure 2-21]. Portable models include Woodhead's Portable

**Figure 2-20.** Hubbell's semiportable GFI. (Courtesy of Hubbell Co.)

**Figure 2-21.** Square D's Quik-Gard GFI/circuit-breaker. (Courtesy of Square D Company.)

Ground-Fault Sentry and Hubbell's GFP-201 four-outlet portable (see Figures 2-22 and 2-23). Hubbell's indoor GFP-115 Mini Portable is attached to a wall receptacle.

Except for bathroom receptacles, plug receptacles are not usually connected to lighting circuits. For that reason lighting circuits have lower priority for receiving GFI protection than do receptacle outlets. When GFIs are incorporated in lighting circuits, emergency lighting must be provided, should the GFI cut off the lights.

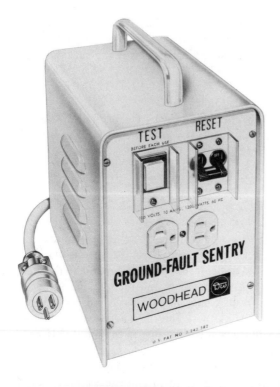

**Figure 2-22.** Woodhead's portable ground fault sentry. (Courtesy of Woodhead Co.)

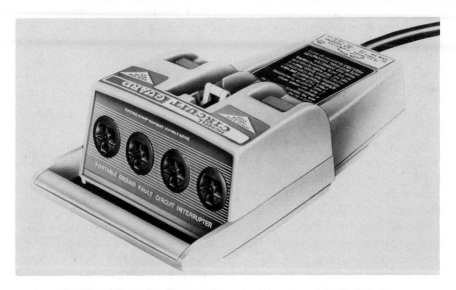

**Figure 2-23.** Hubbell's portable GFI. (Courtesy of Hubbell Co.)

Most GFIs must be operated only on systems that are grounded at the service entrance. However, the Woodhead Portable Ground-Fault Sentry may be used with ungrounded systems. Specific instructions and cautions for installation and use are provided with each GFI.

If a GFI trips, be suspicious of the circuit, not the GFI. DO NOT override the GFI—instead, find and remove the fault.

Several devices are available to check out GFIs, for example, Muska Electric's GFI Trip-Alizer, Daltec Systems Ground Fault Simulator, and Woodhead's GFI Receptacle Tester.

### *Isolation Transformers* (See Figure 2-24A and B.)

Transformerless radio and TV sets may have the chassis connected directly or indirectly to one side of the power line. If so, the chassis may be hot and create a shock hazard. Manufacturers' instruction manuals and Sams Photofacts ® usually point out this problem and recommend the use of an isolation transformer. This transformer has primary and secondary windings completely isolated from each other. The ratio between the windings is 1 to 1, giving an output voltage that is the same as the input voltage. In the better isolation transformers there is a grounded shield between the two windings that prevents insulation failure from destroying the isolation. By isolating the input from the output, the transformer in effect removes the danger of ground fault shocks. To get a shock from equipment powered by an isolation transformer, it's necessary to touch *both* sides of the line. Since this is unlikely, shock protection is excellent.

### *Lockouts*

It is quite simple to ensure that a piece of small equipment cannot shock you during repair—just unplug it from the power receptacle. For larger equipment the power connection may be permanent and the equipment can be deenergized only by throwing the proper disconnect switch or circuit breaker. If the appropriate power switch or circuit breaker is not in direct view of the equipment being repaired, it's possible for someone else to close the switch or circuit breaker inadvertently, not knowing that you may have your hand deep in the guts of a high voltage power supply or other hazardous circuit. To prevent this possibility, you should lock the switch in the OFF position with your own personal padlock (see Figures 2-25A and B). If you can't lock the switch, put a warning sign (see Figure 2-26) on the switch to indicate the danger. Typical words for such a sign are: "Work Being Done On Circuit, Do Not Energize." With a little ingenuity you may be able to

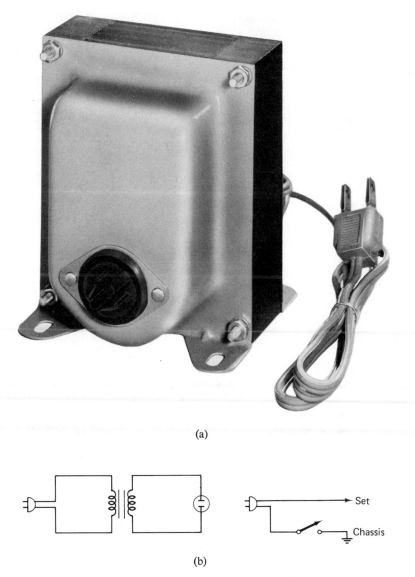

(a)

(b)

**Figure 2-24.** (a) Stancor Model P-6160 isolation transformer. (Courtesy of Essex International, Inc.) (b) Isolation transformer.

devise your own clasps, clamps, or hasps to lock out equipment that cannot ordinarily be locked out. Then, too, it may be well worth the trouble and expense to install disconnect switches for equipment not so fitted now.

After you've locked out a switch box, test the circuit to make sure you've locked out the proper box—the nearest box may not be the right one.

(a)                                                  (b)

**Figure 2-25.** (a) Engineering Development Co.'s lockout device. (Courtesy of Eng. Dev. Co.) (b) Lock-out placed on wire clip over breaker switch. (Courtesy of Osborn Mfg. Corp.)

## *Enclosures*

Permanent or temporary enclosures can provide excellent protection for authorized and unauthorized personnel from accidental contact with uninsulated conductors in electrical equipment. A permanent enclosure may be a simple plastic receiver cabinet, an upright metal transmitter rack, or even an entire room. A temporary enclosure may be a rope or other type of barricade.

Permanent enclosures should be securely fastened (even locked), adequately identified, and equipped with interlocks wherever practical. Metal enclosures must be securely grounded. Metal control shafts that protrude through cabinets also must be grounded. Obviously, no enclosure can be considered safe if contact with live electrical parts can be made from outside the enclosure. And if the interlocks are cheated, enclosures naturally offer very little protection from shock. Open enclosures with energized circuits must never be left unguarded.

Temporary enclosures should be used only when permanent enclosures are impractical or unobtainable. A simple rope stretched between strategically placed chairs may be sufficient to close off a hazardous piece of equipment. Whatever type of barrier is used, it should have good clearance from the equipment. If the barrier is too close, it's conceivable that someone could trip on the barrier and touch a live wire. An ample supply of large (10-by-14 in) DANGER-HIGH VOLTAGE signs should be attached to the barrier

**Figure 2-26.** Typical warning sign for lockout. (Courtesy of Stonehouse Signs, Incorporated.)

and next to the open equipment. DANGER-ELECTRICAL HAZARD
signs should be posted at all entrances to the area containing temporary
enclosures.

### Low-Voltage Circuits

In boilers, tanks, or wet locations, it is sometimes difficult to properly
insulate the worker from ground. In addition, the worker may be perspiring
heavily, which of course increases the possibility of shock. The situation is
made worse by the fact that ground wires in such locations provide no protection against defective extension light cords or broken lamp bulbs with exposed
filaments. A simple solution to this problem is to use low-voltage (6 or 12

volts) lamps and tools powered by an isolation transformer with a grounded static shield between the windings.

### Personal Operating Safety Precautions

Proper insulation, grounding, isolation, and lockouts, plus the use of low voltage and ground fault interrupters can make your shop, laboratory, or transmitter room a much safer place in which to work. However, these precautions may be inadequate to keep you from being shocked, if you have poor or careless work procedures.

Here are some basic procedures to follow in the shop or lab:

1. Keep one hand in your pocket when working on energized circuits.
2. If you can't keep your hand in your pocket, do not touch any metal object with your free hand while holding an electrical tool in the other hand.
3. Don't depend on switches to kill a circuit; pull the plug from the outlet.
4. If you are working on high-voltage circuits, have a buddy standing by to help you in case of shock. Just anybody will not do; your observer must know how to kill the circuit to get you loose and how to give you mouth-to-mouth resuscitation and closed-chest heart massage.
5. Do not wear loose clothing, metallic frame eyeglasses, rings, watches, or other jewelry if you are near energized circuits.
6. Do not use an ordinary lamp as an extension trouble light.
7. If directions or instructions are available, follow them. The guy who wrote the book may know more about the hazards involved than you.
8. When you're mentally or physically fatigued, avoid work on energized circuits.
9. To measure high voltages, deenergize the equipment, discharge appropriate capacitors, attach the meter leads, step back, energize the equipment, and make your readings. Don't go probing with a test lead in your hand.
10. Try to make safety protection equipment fail-safe.
11. Never assume that a circuit is dead. Check it first.
12. Do not rely solely on interlocks unless you are certain that they have disconnected the circuit.

13. Have sufficient illumination to see the smallest parts of the equipment you're working on.
14. Discharge all items that can retain a charge.
15. Short out interlocks at your own risk!
16. Study equipment schematic and instruction manual before you start work.
17. If there is any possibility that the equipment chassis may be hot, connect it to an isolation transformer instead of the usual power outlet.
18. Do not draw arcs either intentionally or accidentally. Technicians on missile work have been nearly blinded by low voltage, high current arcs caused by accidental shorts. Such arcs can also be hazardous in flammable atmospheres.

### *Special Procedures*

#### *High Voltage Wire On Car* [5]

Bear in mind if you touch the car or any of its occupants, you will suffer a serious accident or even death even though the occupants are safe in the car.

If an occupant of the car, in attempting to get out, should touch the ground and the car simultaneously, he will probably be killed.

Call the power company as soon as possible.

Remain calm and speak calmly.
   Don't show excitement or undue concern. People in car may panic.
Stay at least 10 feet from car at all times.
   Don't touch car in contact with a wire or the wire itself with anything held by you.
Tell occupants to stay in car.
   Don't try to judge whether wire is live.
Tell occupants to drive car from contact with wire if it will operate.
   Don't assume that covered wire is insulated. IT IS NOT. Assume any downed wire is dangerous and stay away.
If the car cannot move under its own power and if it is necessary that the occupants leave or be removed from the car, push it from contact with the wire with another car or truck.
   If you use a car or truck to push car in contact with wire, remember the pushing vehicle will be energized upon contact with car with wire

[5]From booklet, "Partnership in Public Safety." Reprinted by permission of the Long Island Lighting Co., Mineola, N.Y.

on it. DON'T leave pushing car or let anyone touch either car until the wire has been cleared by both cars or contact between the two cars has been broken.

If the above suggestions are not feasible: Try wedging between the rear windows of another car a branch of a tree, a length of wood, or some other object so that in protruding from the second vehicle it may be used to snag and remove the wire from the first car, so that its occupants may be removed. If a truck is available, the same operation may be accomplished by wedging the protruding object in some section of the truck.

If you wedge a tree branch, section of wood, or some other object between the rear windows or other part of a second vehicle as a protrusion to snag wire from car, the second vehicle will become energized while the protruding object is contacting the wire or the first car.

DON'T leave or permit anyone to touch second vehicle until the protrusion has been cleared of both the wire and the first car.

DON'T permit anyone to touch any part of the object employed as a protrusion while it is in contact with the wire or the first car.

Only if the situation is critical and there is no alternative should the occupants be advised to jump "clear" while car is still in contact with wire.

*Rescue From Contact With Fallen Wire*[6]

Remember high voltages kill in seconds.

Don't permit a feeling of urgency to influence hasty action which might result in you, too, becoming a victim. Unless only seconds have elapsed since wire fell on victim, it is probably already too late to save his life.

Act with extreme caution.

Don't under any circumstances touch the victim while he is still in contact with wire or touch the wire itself.

Maneuver your car in such a manner as to snag the wire with branch of tree, length of wood, or some other object wedged and protruding from rear window or trunk or some other part of vehicle.

Don't leave your car or permit anyone to touch your car while it or the protruding object is in contact with wire.

Have assistant stand at safe distance until wire has been removed before approaching and removing victim to safe location to administer first aid.

[6]From booklet, "Partnership in Public Safety." Reprinted by permission of the Long Island Lighting Co., Mineola, N.Y.

Don't permit anyone to touch the protruding object while it is in contact with the wire.

## FUNDAMENTAL FIRST AID

Fundamental first aid instruction and training are a necessity for technicians and engineers who work with potentially hazardous electrical equipment. It is especially important for such workers to be trained in mouth-to-mouth resuscitation and closed-chest heart massage. Through these first aid measures it is possible to revive many people from what would otherwise be a fatal shock.

The following paragraphs should not be considered an instruction course for cardiopulmonary resuscitation (the combination of mouth-to-mouth resuscitation and closed-chest heart massage). Classroom training using special manikins to practice on is essential for the prospective first-aider in electric shock cases. This is particularly true for closed chest massage, which may cause cracked ribs or a punctured lung if not done properly.

Classes on cardiopulmonary resuscitation (CPR) are given by several groups, including some YMCA's, electric utilties, and military groups. Look for them in your area.

The intent of the following paragraphs is not to show you how to do CPR but to show you how relatively simple CPR is. These paragraphs have been extracted from information supplied by the Edison Electric Institute, the American National Red Cross, and the American Heart Association.

In some cases the victim of a severe electric shock may be thrown from a hot line or equipment because of muscle spasms, while at other times the victim may remain frozen to the point of contact. To free such a victim without becoming a victim himself, a rescuer must be extremely careful. The safest way, of course, is to open the proper switch or to use well-insulated tools to cut the power line. If this is not possible, try to find several rubber mats or a *dry* rope to pull the victim loose. Freeing the victim this way can be a very tricky procedure and requires the utmost caution. Obviously the farther away you, as the rescuer, are from the victim the safer you will be.

Once the victim has been removed from the circuit and both you and the victim are at a safe distance from the hot line, you must start CPR immediately—seconds count. Within four to six minutes the brain will suffer irreparable damage unless CPR is started (see Figure 2-27). Therefore, there is no time to drag the victim to a more comfortable location or to call for help, or to even loosen his clothes.

Naturally, if the victim is conscious, there is no need for CPR. If he is unconscious, the first step is to observe his chest, listen at his nose and mouth, and feel for the movement of air. If none of these indications of breathing

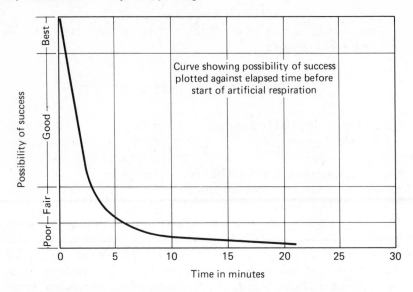

**Figure 2-27.** Possibility of success vs elapsed time before start of artificial respiration. (Courtesy of Edison Electric Institute.)

are present, give the victim six quick breaths using the following procedure (see Figure 2-28):

1. If foreign matter is visible in the mouth, wipe it out quickly with your fingers, wrapped in a cloth, if possible.

2. Tilt the victim's head backward so that his chin is pointing upward. This is accomplished by placing one hand under the victim's neck and lifting, while the other hand is placed on his forehead and pressing. This procedure should provide an open airway by moving the tongue away from the back of the throat.

3. Maintain the backward head-tilt position and, to prevent leakage of air, pinch the victim's nostrils with the fingers of the hand that is pressing on the forehead.

   Open your mouth wide; take a deep breath; and seal your mouth tightly around the victim's mouth with a wide-open circle and blow into his mouth. If the airway is clear, only moderate resistance to the blowing effort will be felt.

   If you are not getting air exchange, check to see if there is a foreign body in the back of the mouth obstructing the air passages. Reposition the head and resume the blowing effort.

4. Watch the victim's chest, and when you see it rise, stop inflation, raise your mouth, turn your head to the side, and listen for exhalation. Watch the chest to see that it falls.

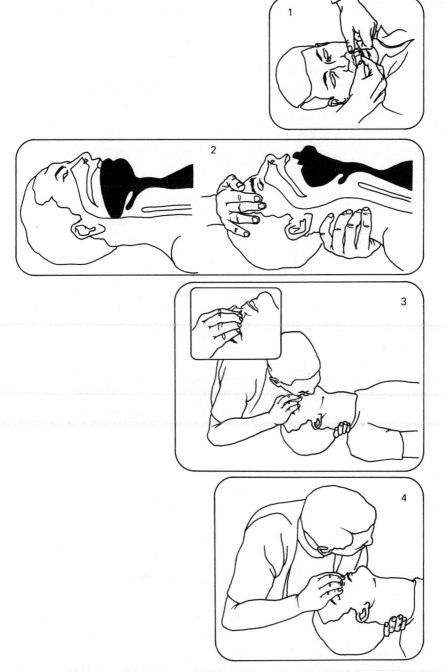

**Figure 2-28.** Mouth-to-mouth resuscitation. (Reprinted with permission of the American Red Cross.)

When his exhalation is finished, repeat the blowing cycle, for a total of 6 breaths. Volume is important. Start at a high rate and then provide at least one breath every 5 seconds for adults.

Steps 1 through 4 above, provided by the American National Red Cross, constitute the familiar mouth-to-mouth resuscitation technique. It is considered a simple procedure and can be applied even without manikin practice.

Before continuing with mouth-to-mouth resuscitation, it's important to stop long enough to determine if the victim's heart is beating. To do this feel for a pulse alongside the victim's Adam's apple. If you can't detect a pulse, raise one of the victim's eyelids and observe the pupil. If it is enlarged and does not narrow down in response to bright light, then the heart has stopped or is fibrillating. In either case, mouth-to-mouth resuscitation alone will probably not be enough to revive the victim; it must be combined with external heart compression.

> CAUTION: The American Heart Association warns that heart compression (massage) should be done only by well-trained individuals.

*External Heart Compression.*[7] After the victim has been given six breaths by the preceding procedure:

1. Place the heel of one hand on the lower third of the victim's breastbone (sternum) and the other hand on top of the first (see Figure 2-29). To find this spot, put a finger on the bottom end of the breastbone and place the hand alongside it. According to the American Heart Association, "the pressure point is on the lower half of the breast plate just

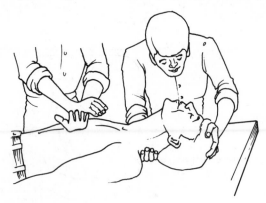

**Figure 2-29.** External heart compression. (Courtesy of Edison Electric Institute.)

[7] *Resuscitation Manual,* Edison Electric Institute. Reprinted by permission of Edison Electric Institute.

above its soft lower end where it joins the abdomen. Pressure must be applied here and nowhere else if the method is to be effective and if breaking of the ribs and damage to organs inside the body are to be avoided. This is extremely important."

2. With the fingers extended so that no pressure is applied to the ribs, press down firmly and quickly, so that the breastbone is depressed about 1½ to 2 inches.

3. If you're working alone, interrupt the heart compressions about every 15 or 20 strokes to give two or three breaths of air to the victim. If another rescuer is available, one of you should apply heart compressions at 60 cycles per minute while the other gives mouth-to-mouth resuscitation at 12 cycles per minute.

*When to Stop.* Continue resuscitation until the victim revives, medical help arrives, or rigor mortis sets in. If you are applying heart compression, do not use a "pressure-cycled" mechanical resuscitator for ventilating the lungs.

*Do Not:*
Give liquids to an unconscious person.
Give heart compression to a victim who has a pulse, broken ribs, or pupils that do not remain widely dilated.

# 3

# STATIC ELECTRICITY

For most of us, static electricity (Figure 3-1) is not really a hazard; it's merely a nuisance, a minor shock, that we encounter on cold, dry winter days as we shuffle across a carpeted floor or as we slide across the plastic seat covers in our automobiles. However uncomfortable or annoying such jolts may be, they are not going to electrocute you. Aside from lightning, which is a form of static electricity, there is no record of anyone being killed by static electricity shock.

Nevertheless, under certain common conditions static electricity presents three distinct hazards:

1. It can zap semiconductors, particularly some types of MOSFETs, destroying them in the process. Very thin oxide layers separate metallic contacts in MOSFETs; if potentials of roughly 100 volts are applied, these layers break down. Because of this problem, special precautions must be taken in the manufacture, shipment, and use of some MOSFETs. This breakdown problem affects not only transistorized computers, but also transistorized stereo or television sets.

2. It can startle you into an accident, since it is natural to jump or jerk away from a static electricity shock. Chances are that no great harm will be done, but if you're in a precarious position, on a ladder, for instance, or around dangerous, unguarded moving machinery, your involuntary reflex may turn a minor incident into a major accident. It should also be noted that the National Safety Council advises persons with weak hearts to avoid operating electrostatic painting equipment because minor electrical shock might create heart complications.

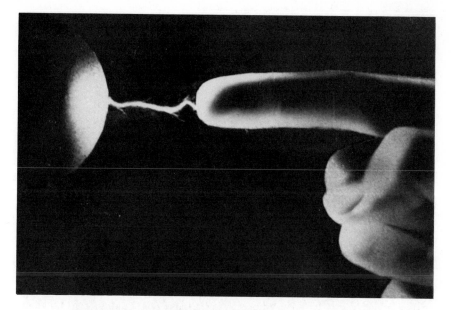

**Figure 3-1.** Typical static electricity discharge. (Courtesy of Custom Materials, Inc.)

3. It can cause sudden, devastating fires and explosions around munitions, missile propellants, hospital operating rooms, solvents, fuels, and even dust (see Figure 3-2). Although precise figures may never be known, each year people are killed and thousands of dollars of damage results from fires and explosions that are probably caused by static electricity; also many fires of "unknown origin."

Flammable liquids used in dry cleaning, printing, and other activities may be ignited by static electricity during handling or storage. Petroleum products such as kerosene and fuel oil can generate and retain large charges when they are pumped through pipes; when these charged products are placed in tanks, the vapor space above the fuel may ignite from the charge. A static spark or discharge that is too small to be seen or felt may contain more than enough energy to ignite some explosive vapors.

In hospital operating rooms using explosive anesthetics the danger of static electricity has long been recognized. Patients undergoing surgery have died from such explosions.

Combustible dusts—such as grain dust, organic dust, and metallic dust—can easily acquire electrical charges. In the process of settling upon insulated surfaces these dusts may cause appreciable accumulation of static electricity, resulting in disastrous sparks. It takes only a low intensity spark to ignite combustible dust.

**Figure 3-2.** Static electricity was the probable cause of the accidental ignition of a Delta rocket motor at Cape Canaveral. (Courtesy of of National Aeronautics and Space Administration.)

Tank trucks and other rubber-tired vehicles including automobiles can generate static electricity through the continual contact and separation of the tires with the roadway, especially when the tires are dry. This potential source of ignition is a matter of real concern when rubber-tired vehicles are used to transport flammable liquids such as gasoline.

In plants that manufacture and deal with munitions and explosives, stringent measures must be taken to avoid sparking from static electricity. Missile propellants likewise must be guarded against accident. The simple removal of a polyethylene sheet from a missile a few years ago at Cape Kennedy was sufficient to generate a static discharge that ignited the missile.

Even when static electricity is not hazardous, it can be a nuisance—in printing plants and graphic arts departments, for example, static electricity slows production and interferes with smooth operation by causing paper to stick together.

Truly, static electricity, which has been observed for hundreds of years, has now become a space age gremlin, harmless at times, deadly on occasion.

In this chapter we will consider the sources of static electricity, how it can be detected and measured, and how it can be controlled. Lightning will be discussed in Chapter 4. Much of the following material was derived from *Static Electricity*, a publication of the U.S. Department of Labor, and from material supplied by Custom Materials, Inc.

## GENERATION

Static electricity may be built up at any time there is motion, or friction, between two objects, especially if the objects are made of dissimilar materials, or when there is electrostatic induction. When the two objects are in contact, electrons move from one object to the other. If one or both objects are nonconducting when the objects separate, electrical neutrality cannot be immediately reestablished. As a result, one object accumulates a negative charge and the other object accumulates an equal and opposite positive charge. If the surface has a deficiency or excess of only one electron in 100,000 atoms, it is considered to be strongly charged.

The accumulated charge, which is trapped, or prevented from escaping, has no place to go. Because it does not move, it is called *static* electricity. The more perfect the insulator, the longer the object will retain the accumulation. If the charge leaks off rapidly, there will be no problem. However, if it accumulates, at some point this accumulation will become sufficient to jump to a nearby object with lower potential, forming a spark in the process.

The spark may be too small to be seen or it may be as long as eight inches, according to the National Fire Protection Association. Regardless of its size, this spark discharge is the major hazard of static electricity.

The energy in the spark is given by:

$$\text{energy (joules)} = \tfrac{1}{2}CV^2$$

where $C$ is capacitance in farads and $V$ is the voltage. For a person, $C$ is considered to be 100 pF and $V$ varies with the situation. While sitting on a varnished wooden stool, for example, you may generate up to 20,000 volts on your body. A simple walk across the floor may allow you to accumulate up to 50,000 volts if humidity is low. Even nearby static charges may induce high static voltages on you. Thus, for the energy equation, we can assume a $V$ of 50,000 volts. Under these conditions, a person may cause a spark of 0.01 joule, which is more than enough energy to ignite certain vapors (see Table 3-1).

An excellent way to deliberately generate static electricity is to move plastics or other poor conductors rapidly over rollers or pulleys. This is the method used to produce high voltage discharges in the familiar Van de Graaff generators (see Figure 3-3). Even with many thousands of volts being produced by these generators, it should be noted that they are relatively safe when used according to directions.

**TABLE 3-1** *The following chart is a handy reference to the explosive range of various liquids and gases and the* **minimum** *ignition energy required to explode them. As a rule of thumb, the human body can easily build up a static charge energy of .02 joules.*

|  | Explosive Range (% concentration by volume) | Minimum Ignition Energy (in Joules) |
|---|---|---|
| ACETONE | 2.6 –12.8 | .0006 |
| BENZENE | 1.4 – 7.1 | .0005 |
| BUTANE | 1.9 – 8.5 | .00064 |
| CARBON DISULPHIDE | 1.25–50.0 | .00015 |
| CYCLOPROPANE | 2.4 –10.4 | .00018 |
| ETHER (Diethyl) | 1.85–36.5 | .00045 |
| ALCOHOL (Ethyl) | 3.3 –19.0 | .00065 |
| GASOLINE | 1.4 – 7.6 | .001 |
| HYDROGEN | 4.0 –75.0 | .00002 |
| METHANE | 5.3 –15.0 | .00096 |
| METHYL ALCOHOL | 6.7 –36.0 | .0005 |
| PROPANE | 2.2 – 9.5 | .00036 |

Reference: Federal Bureau of Mines, National Fire Protection Association (Reprinted by permission of Custom Materials, Inc.)

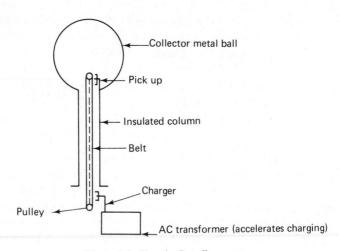

**Figure 3-3.** Van de Graaff generator.

In his book, *Electrostatics: Exploring, Controlling & Using Static Electricity*, (Doubleday,N.Y., 1968), Professor A. D. Moore points out that thousands of Van de Graaff generators have been in use for years in high school and college classrooms without any significant mishap. The very act of sparking, he notes, largely discharges such generators and therefore they cannot continuously pour energy into you if you are holding on to them.

If, however, you foolishly decide to add one or more capacitors to the generator, you may set yourself (or someone else) up for a fatal shock. The amount of energy that a capacitor will store up can also be found by the equation

$$\text{energy} = \tfrac{1}{2}CV^2$$

It is important to realize that a discharge into the body exceeding 10 joules can be hazardous to life, and that only $\tfrac{1}{4}$ joule can cause a severe shock. (The National Fire Protection Association notes that 1 joule is equivalent to being hit on the jaw by a $4\tfrac{1}{2}$-pound sledge hammer that has traveled a yard in 1 second.) The following table gives typical combinations of voltages and capacitances that can be fatal:

| Microfarads: | 0.002 | 0.20 | 20.00 | 80.00 | .320.00 | 2000.00 |
|---|---|---|---|---|---|---|
| Kilovolts: | 100.000 | 10.00 | 1.00 | 0.50 | 0.25 | 0.10 |

Thus, any capacitance used with static generators must be considerably less than these values if safety is to be maintained. If you use capacitors, discharge them immediately when not in use and keep them shorted to prevent them from recharging (even if they are out of the circuit).

## DETECTION AND MEASUREMENT

In many situations test instruments are not needed to detect the presence of static electricity: you can hear its crackle, see clothes clinging, or feel it move the hair on the back of your arm. In fact, static voltages as low as 1,500 volts can be detected with your little finger, according to the Bureau of Mines, if you bring it slowly and delicately to within a few thousandths of an inch of the charged object. For more sensitive detection and for measurement, you should resort to electroscopes, neon glow tubes, electrostatic voltmeters, or vacuum tube electrometers. (If it's necessary to use battery-operated or line-operated static measuring equipment in atmospheres containing explosive vapors or dust clouds, be very, very careful: sparks from the equipment may ignite the vapors.)

*Gold Leaf Electroscope.* A gold leaf electroscope (see Figure 3-4) is the simplest device for detecting and measuring static electricity. Basically, it consists of a thin gold leaf fastened to a metal rod. The rod is mounted in a box and is insulated from it by an amber plug. On the other end of the rod, outside the box, is a metal knob which is used to touch charged surfaces. When the knob touches a charged surface, the gold leaf is repelled by the presence of like charges on the leaf and the post, and moves across a calibrated scale. The scale indication, which is roughly proportional to the relative potential, can be observed through a window in the box.

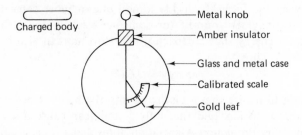

Charged body

Metal knob

Amber insulator

Glass and metal case

Calibrated scale

Gold leaf

**Figure 3-4.** Gold leaf electroscope.

*Neon Glow Tubes.* Glow tubes similar to those used to check automobile spark plugs are useful in determining the presence of static electricity; however, they do not indicate the amount of the accumulation.

*Electrostatic Voltmeters or Electrometers.* Electrostatic voltmeters or electrometers operate by measuring the movement between fixed and mobile vanes on which a voltage is impressed. When some part of the instrument or probe is held a short distance away from the surface to be tested, the charge on the surface deflects a spring-loaded, movable vane. A needle attached to this vane indicates the amount of the charge.

*Vacuum Tube Electrometers.* The vacuum tube electrometer (see Figure 3-5) uses a simple radio tube with a heated filament as a source of electrons and a plate upon which a strong positive charge is impressed. Static charges impressed on a grid placed between the filament and the plate will modify

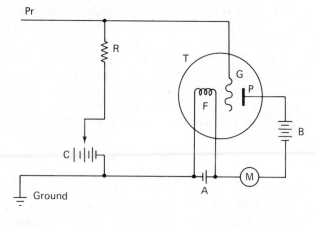

| | | |
|---|---|---|
| T — Vacuum tube | M — Meter | C — C battery to adjust |
| P — Plate | Pr — Probe | grid-plate circuit current |
| F — Filament | R — Resistance to | B — B battery to impress a |
| G — Grid | control grid | charge on plate |
| | leakage | A — A battery to heat filament |

**Figure 3-5.** Vacuum tube electroscope.

the current flowing from the filament to the plate. If a positive charge is placed upon the grid, the current will increase; if a negative charge is impressed upon the grid, the current will decrease. A meter in the plate circuit indicates these changes in current.

For accurate measurements, make sure that:

1. You don't become charged yourself. A charge on your body could throw the instrument indication off.
2. You take all measurements under the same conditions.
3. You check readings against adjacent grounded equipment.

*Static Meter.* Custom Materials' Static Meter (Figure 3-6), a typical commercial model, measures 0 to 200,000 volts. A special high resistance measuring element in the meter couples to the atmosphere by means of a radioactive beta emitter (tritium).

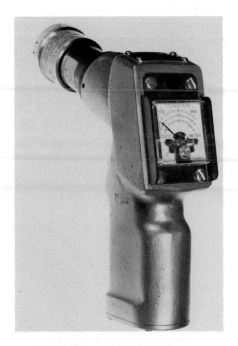

**Figure 3-6.** Custom Materials static meter. (Courtesy of Custom Materials, Inc.)

## CONTROL

In many situations it's impossible not to generate static electricity; however, generation is not the hazard—the danger lies in the accumulation and subsequent discharge of this static charge.

The way to control static electricity is to provide a path by which separated charges may recombine harmlessly before a spark can occur. Three common ways for doing this are bonding and grounding, ionization of the surrounding media, and humidification.

*Bonding and Grounding.* When two conducting bodies are connected with a conducting wire, they are said to be *bonded*. As long as they are bonded, they will be at the same electrostatic potential and sparks will not jump between the two.

However, bonded objects may have a different potential from other objects and from ground. In such cases, sparking may occur. To prevent this, it is necessary only to ground the bonded objects. To ensure that the bond or ground wire has adequate mechanical strength, it should be No. 8 or No. 10 AWG, even though smaller wire would be sufficient for the small currents involved in static electricity. (In munitions areas, No. 8 AWG wire is required.) Either insulated or noninsulated wire may be used. If insulated wire is used, it should be checked more frequently for electrical continuity, since breaks in the wire cannot be seen. In either case it's important to use corrosion-resistant wire.

Ground connections must be made firmly and securely to ensure a low resistance ground (Figure 3-7) Some authorities advise not using a power circuit ground as a static ground.

In particularly hazardous areas, such as hospital operating rooms and munitions areas, it's necessary to bond and ground personnel as well as equipment (Figure 3-8). One method for accomplishing this is to use a con-

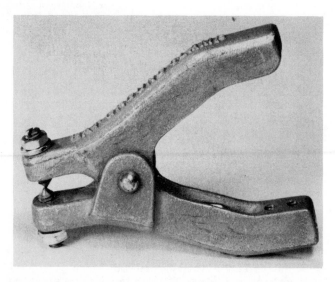

Figure 3-7. Static electricity grounding clamp. (Courtesy of Stewart R. Browne Manufacturing Co.)

Personnel grounding

How static build-up can cause
static hazard

Static charge build-up on this man cannot
follow to the ground. There is no
conductive path.

The proper set-up to eliminate
static hazard

The "Velostat" boots or heel grounding
protectors provide a conductive path for
the static charge to be drained instantaneously
to ground. There is no static build-up to cause
a spark.

"Velostat" boots or heel
grounding protectors.

"Velostat" floor mat.

Ground

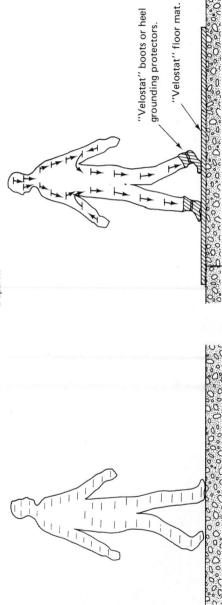

**Figure 3-8.** Static charge build-up and prevention. (Courtesy of
Custom Materials, Inc.)

ductive floor, that is, a floor with a resistance less than 1 megohm per 3 feet. In a hospital operating room, for example, the resistance may vary from 25,000 ohms to 1 megohm. Although this may seem to be a high resistance, it is low enough to drain off static charges, yet high enough to minimize conventional electrical shock hazards. By comparison, an ordinary concrete floor has a resistance 100 times greater than such a conductive floor.

In conjunction with conductive floors, special personnel grounding devices and clothing must be worn to minimize the possibility of sparking. These include conductive shoes (or grounding devices attached to regular shoes—Figure 3-9), clothes made with metal fibers, and wrist straps (Figure 3-10) attached to a ground ribbon, which is connected to a work station ground.

Underclothes and outer garments made of synthetic fibers, wool, and silk should be avoided; instead, wear cotton or linen garments, including socks.

Custom Materials, Inc. makes an electrically conductive plastic (under the trademark "Velostat"), which is made into drum liners and bags, containers, gloves, aprons, tarpaulins, etc. Use of this material ensures bonding and grounding of items that otherwise could not be grounded.

Other applications of bonding and grounding include:

1. Use of conductive belts, made with interwoven wires or conductive additives, in machinery using moving belts.

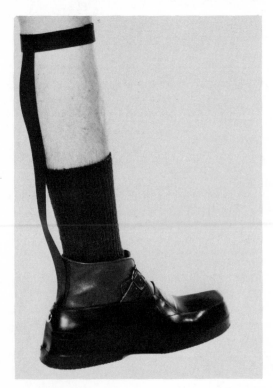

**Figure 3-9.** Velostat conductive shoe with leg strap. (Courtesy of Custom Materials, Inc.)

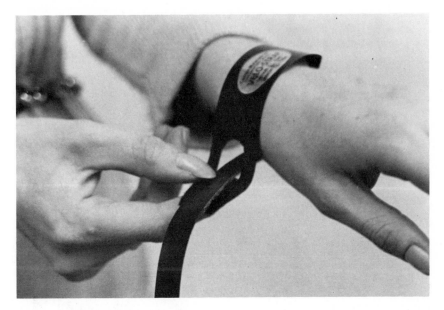

**Figure 3-10.** Velostat conductive wrist strap. (Courtesy of Custom Materials, Inc.)

2. Use of semiconductive tires on rubber-tired vehicles carrying flammable liquids.
3. Bonding of tank trucks to unloading docks or to the inlet pipes of loading stations *before* transferral of flammable liquids.
4. Bonding and grounding all containers used in pouring and transferring flammable liquids (including paint thinners and gasoline—see Figure 3-11).

*Ionization.* When grounding is not feasible, the air in immediate contact with a charged body is ionized to create a conductive path through which the static electricity can be discharged to ground. Static combs or brushes, high voltage eliminators, electronic static bars, and radioactive static eliminators make use of ionization to allow charges to be safely dissipated to ground.

A *static comb* or *brush* consists of a metal bar with rows of needle points or a metal wire wrapped with metallic tinsel. When the comb is brought near an insulated charged object, the object will induce an ionizing charge on the comb. Because of the charge on the comb, the air surrounding the points of the comb will be ionized and will then provide a path to ground for the static electricity.

By impressing a high voltage on the comb, a high-voltage eliminator is created, which is more efficient than the conventional static comb (Figure

Conditions under which many static spark initiated fires and explosions have taken place

Person approaching the solvent drum to draw off solvent is highly charged with static electricity due to the motion of walking.

Solvent drum usually possesses a minimum of electrical charge if it stands a long period of time without motion. Static spark can take place anywhere between the person carrying draw off container and the drum hardware.

Flammable solvent drum

100% safe conditions against static spark hazard when working with flammable solvents. Use electrically conductive containers for storing or drawing of solvents. Use conductive plastics or metals. Avoid use of glass containers.

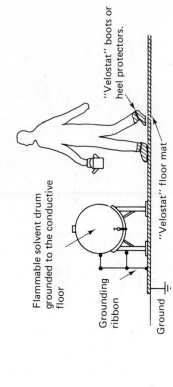

Flammable solvent drum grounded to the conductive floor

Grounding ribbon

Ground

"Velostat" floor mat

"Velostat" boots or heel protectors.

**Figure 3-11.** Static build-up and protection of solvent drums. (Courtesy of Custom Materials, Inc.)

62

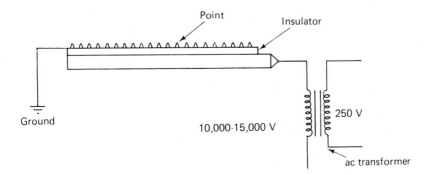

**Figure 3-12.** High-voltage eliminator.

3-12). However, because of the high voltage, steps must be taken to prevent electric shock; high-voltage eliminators should not be used in atmospheres containing flammable vapors.

The *electronic static bar* is an improved version of the high-voltage eliminator that maintains electronically the high voltage at very low currents. In contrast to the high-voltage eliminator, the static bar can safely be used in flammable atmospheres; it also has a lower shock potential.

*Radioactive static eliminators* employ a radioactive isotope such as radium, polonium, or strontium 90 to bombard air molecules and produce numerous ions. Special techniques are used to keep the radioactive material from escaping into the environment and to reduce radiation exposure to personnel. Nevertheless, such eliminators must be used very carefully and must be checked periodically by trained personnel to detect radiation hazards.

*Humidification.* When static electricity causes sheets of paper or layers of cloth to stick together, or vapors to explode, humidification may be the answer. When the humidity is increased, the surface conductivity of nonconductive materials increases, which enables static accumulations to leak off to ground more rapidly. When the humidity decreases, as in cold, dry months, static tends to accumulate more. (Note that the rate of generation is the same in the summer as in the winter.)

Antistatic chemicals work best when the humidity is high. They pick up moisture from the air and thereby make the surface of the material they are applied to very slightly conductive. This allows charges built up on the material to flow off. Note that antistatic materials can be applied to materials, such as sheets of polyethylene, that normally are insulators and hold charges even in high humidity unless they are so treated.

To achieve the best static control, it's necessary to have a relative humidity of 60 to 70%, which can best be determined by the use of wet and dry bulb thermometers; dial-type hygrometers may not be sufficiently accurate.

Humidification is not widely used because high humidity can obviously be uncomfortable to workers and it can sometimes damage equipment and materials.

## SPECIAL PROBLEMS

*Hospital Operating Rooms.* Because of the explosive anesthetics used in hospital operating rooms, it is imperative that strict static electricity control measures be taken to avoid disaster. The most important of these measures are the use of conductive flooring, the grounding of all equipment and personnel in the explosive area, and the replacement of insulative cloth and plastics with either antistatic materials or conductive materials.

Ordinary plastics, paper, and cloth are not normally safe to use because they cannot be grounded by conventional means. Antistatic chemicals can make them safe provided that: (1) the chemicals have been stored properly but have not aged, (2) proper humidity is maintained, and (3) the chemicals are tested if there is any question of their effectiveness. Conductive materials, on the other hand, do not have these problems of humidity, aging, or exposure to sunlight.

*Munitions and Explosives Manufacturing and Handling.* Electrically-initiated explosive devices such as primers, squibs, blasting caps, and dimple motors can also, unfortunately, be ignited by static electricity. Because of this, it is extremely important that all personnel involved in manufacturing and handling such devices (Figure 3-13) avoid any unplanned discharge of static electricity. Proper precautions include use of separate ground systems with low resistance (10 or 25 ohms); conductive floors, shoes, and table tops (Figure 3-14); and frequent, periodic inspections and tests of static ground systems and equipment. For further information, refer to Custom Materials report *Electrical Grounding Concepts* and Air Force manual AFM 127-100.

*Semiconductor Manufacture and Handling.* The General Electric company recommends the following procedure for handling metal-oxide semiconductor field effect devices:

1. Before handling these devices, the user or the equipment should momentarily contact a metal object at electrical ground potential so that any static charges will be removed.
2. The same electrical ground potential should be applied to any lead-shorting and device-shrouding materials and to the device case when possible.
3. Soldering-iron tips should be electrically grounded before soldering any wire or metal objects that are directly or indirectly connected to the device.

Serious static spark hazard exists when arming rocket motors

1. Personnel working on arming motors or devices must be properly grounded electrically.
2. Care must be taken to make certain metal housings for squibs and igniters are grounded before positioning.
3. Always check to see that lead wires of squibs or igniters are twisted together and case grounded.

Working under these conditions provides maximum safety. Further operations to connect lead wires, etc., commence with zero static accumulation and zero static charge potential.

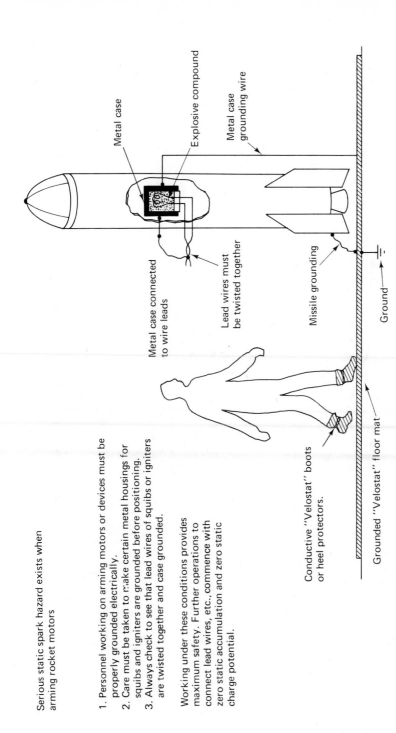

Metal case

Explosive compound

Metal case grounding wire

Metal case connected to wire leads

Lead wires must be twisted together

Missile grounding

Conductive "Velostat" boots or heel protectors.

Grounded "Velostat" floor mat

Ground

**Figure 3-13.** Grounding precautions with propellants. (Courtesy of Custom Materials, Inc.)

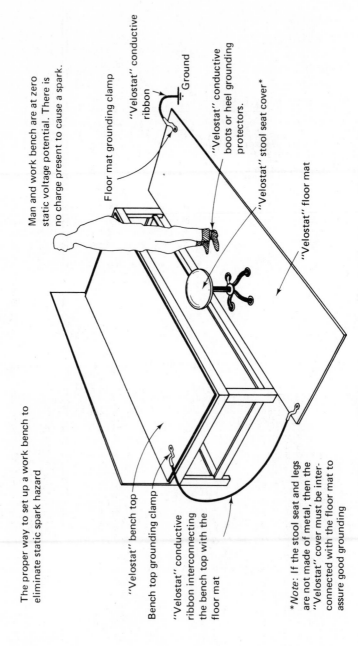

The proper way to set up a work bench to eliminate static spark hazard

Man and work bench are at zero static voltage potential. There is no charge present to cause a spark.

Floor mat grounding clamp

"Velostat" conductive ribbon

Ground

"Velostat" conductive boots or heel grounding protectors.

"Velostat" stool seat cover*

"Velostat" floor mat

"Velostat" bench top

Bench top grounding clamp

"Velostat" conductive ribbon interconnecting the bench top with the floor mat

*Note: If the stool seat and legs are not made of metal, then the "Velostat" cover must be interconnected with the floor mat to assure good grounding

**Figure 3-14** Conductive work station. (Courtesy of Custom Materials, Inc.)

66

4. In general, electrical equipment metal cabinets and the chassis that will contain these devices should be electrically grounded.

5. It is recommended that any lead-shorting mechanisms not be removed until insertion or assembly into equipment. When these devices are being transported or stored at the user's facility, the device leads should be electrically shorted together. A suggested method for various packages is as follows:

a. TO 5 to TO 18 package—retain the shipping shorting mechanism or place a metal washer, metal foil, or conductive foam over the leads and ensure that electrical contact is made to all leads.

Custom Materials offers the following additional suggestions for handling MOSFETs:

1. Provide properly grounded conductive table tops, floor mats, and chair or stool covers.

2. Provide personnel grounding to the extent that grounding wrist straps and heel protectors are used to achieve the required electrical ground.

3. In the handling for assembly into printed circuit cards, transportation of parts and cards, storage, card test, assembly into systems and packaging of MOSFET devices, use electrically conductive Velofoam (Figure 3-15).

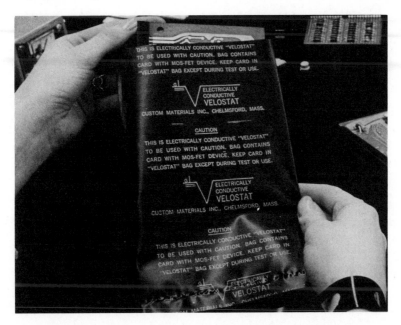

**Figure 3-15.** Velostat conductive bag for transporting MOSFET devices. (Courtesy of Custom Materials, Inc.)

4. Conductive aprons are available to provide conductive and grounded body shield to further protect from static buildup on clothing in vicinity of work bench area.
5. Make it mandatory that personnel avoid the use of synthetic fiber clothing and that they use cotton as much as possible.

When using conductive materials to avoid static discharge, be particularly careful to prevent electric shock from conventional sources as the extensive use of conductive material increases the possibility of shock.

# 4

# LIGHTNING

From a distance, lightning is an awesome yet beautiful display of electrical fireworks. In less than half a second, a 100-million volt bolt can streak from cloud to cloud, or from earth to cloud, leaving a distant rumble behind (see Figure 4-1). Peak current may range up to 200,000 amps, which does not seem so much at a distance. But up close, this discharge of power can be frightening as well as dangerous, as it can electrocute any person in the way or crush any structure that becomes part of the circuit.

At any given moment, it's estimated that 1,800 thunderstorms, the major sources of lightning, are in progress around the world and that lightning strikes the earth 100 times each second. In the United States alone, lightning kills about 150 people each year and injures another 250. (Note that these figures may be too conservative; some authorities say that 600 are killed and 1,500 injured.) Property damage and other losses from lightning are estimated to be more than $200 million each year.

As compared to other dangers, such as automobile accidents, the danger of lightning may seem unlikely, perhaps even exaggerated. But it must be remembered that lightning is a greater killer than hurricanes or tornados. And electronic communication personnel are much more exposed to lightning than is the average person—many communication facilities are in rural areas and on the highest peaks, prime target areas for lightning.

In moderately flat areas of the northeast, according to Professor Martin Uman, the expected number of lightning strokes per year is about 1 for a 300-ft structure, 3 for a 600-ft structure, and 20 for a 1,200-ft structure. Obviously, many radio and TV towers are excellent targets. However, with proper lightning protection, the danger to equipment and personnel is slight.

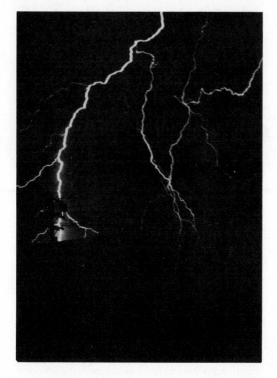

**Figure 4-1.** Lightning. (Courtesy of Thompson Lightning Protection, Inc.)

— Whether it's a barn, a house (Figure 4-2), or communication facility, the standard approach to lightning protection is to provide a planned, safe path for lightning strokes to travel on from the highest points of the structure to ground.

On a building, for example, pointed metal rods (called *air terminals*) about 10 to 24 inches long are mounted on high points of the building (Figure 4-3). (Instead of terminals, some authorities recommend wire conductor along the peak of the roof.) Heavy electrical conductors, generally No. 6 AWG copper or aluminum, are run from the terminals to a low resistance ground. All metal structures and pipes in the building are then bonded to this system. Finally, arresters, protectors, and grounding blocks are attached to incoming power and communications lines that cannot be grounded (see Figures 4-4 through 4-6).

(Note: Even if you don't have a lightning protection system for your house, make sure the masts of roof-top antennas are grounded to comply with the National Electrical Code.)

One of the newest approaches to lightning protection is to *prevent* lightning strokes rather than merely to channel them to ground after they've already been formed (see Figures 4-7 and 4-8). Lightning Elimination Associates (LEA)'s Dissipation Array is said to reduce the space charge between

Lightning protection points for a house: (1) terminals spaced a minimum of 20 feet apart along ridges and within 2 feet of ridge ends; (2) downlead conductors; (3) at least two grounds, at least 10 feet deep, for house — additional grounds for clotheslines, etc.; (4) roof projections such as ornaments tied into conductor system; (5) protection for tree within 10 feet of house — connect to house grounding; (6) at least two terminals on chimneys; (7) dormers rodded; (8) arrester on antenna — connect to main conductor; (9) tie-in to conductor system of gutter within 6 feet of conductor; (10) arrester on overhead power lines.

**Figure 4-2.** Lightning protection points for a house. (From *Lightning Protection for the Farm*, Farmers' Bulletin No. 2136, 1968, U.S. Department of Agriculture.)

the clouds and the protected facility to well below the discharge potential by bleeding off the energy slowly over a period of time. As an example of its effectiveness, a 1,200-ft Air Force tower in Florida received an average of 100 strikes per year before installation of the LEA Array; now it receives none. The Array must be custom designed to fit each installation.

Aside from lightning prevention and lightning control techniques, there are some well-established techniques for protecting yourself from lightning. The National Oceanic and Atmospheric Administration gives these lightning safety rules when lightning threatens:

1. Stay indoors, and don't venture outside, unless absolutely necessary.
2. Stay away from open doors and windows, fireplaces, radiators, stoves, metal pipes, sinks, and plug-in electrical appliances.
3. Don't use plug-in electrical equipment like hair dryers, electric tooth brushes, or electric razors during the storm.

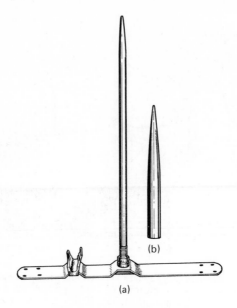

Common types of air terminals
are (a) screw-in mounting and
(b) screw-on mounting.

**Figure 4-3.** Air terminals. (From *Lightning Protection for the Farm*, Farmers' Bulletin No. 2136, 1968, U.S. Department of Agriculture.)

**Figure 4-4.** Secondary power arrester mounted on side of fuse box. (Courtesy of Dale Electronics.)

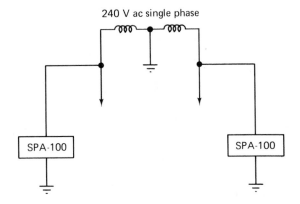

240 V ac single phase

SPA-100          SPA-100

**Figure 4-5.** Wiring diagram for Dale SPA arrester. (Courtesy of Dale Electronics.)

**Figure 4-6.** Grounding block for CATV-MATV lines. (Courtesy of AVA Electronics Corp.)

4. Don't use the telephone during the storm—lightning may strike telephone lines outside.

5. Don't work on fences, telephone or power lines, pipelines, or structural steel fabrication.

6. Don't use metal objects like fishing rods and golf clubs. Golfers wearing cleated shoes are particularly good lightning rods.

7. Don't handle flammable materials in open containers.

8. Stop tractor work, especially when the tractor is pulling metal equipment, and dismount. Tractors and other implements in metallic contact with the ground are often struck by lightning.

9. Get out of the water and off small boats.

● The physical situation

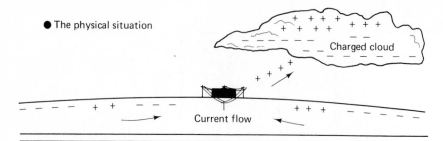

Charged cloud

Current flow

● The equivalent circuit

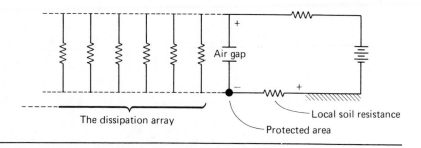

Air gap

The dissipation array

Local soil resistance

Protected area

**Figure 4-7.** Principles of operation of LEA's lightning dissipation array. (Courtesy of Lightning Elimination Associates.)

**Figure 4-8.** LEA's lightning dissipation array. (Courtesy of Lightning Elimination Associates.)

10. Stay in your automobile if you are traveling. Automobiles offer excellent lightning protection.

11. Seek shelter in buildings. If no buildings are available, your best protection is a cave, ditch, canyon, or under head-high clumps of trees in open forest glades.

12. When there is no shelter, avoid the highest object in the area. If only isolated trees are nearby, your best protection is to crouch in the open, keeping twice as far away from isolated trees as the trees are high.

13. Avoid hill tops, open spaces, wire fences, metal clothes lines, exposed sheds, and any electrically conductive elevated objects.

14. When you feel an electrical charge—if your hair stands on end or your skin tingles—lightning may be about to strike you. Drop to the ground immediately.

    NASA's Kennedy Space Center's Lightning Study Team recommends you seek shelter when the time lag between lightning flash and the clap of the thunder is 25 seconds (that is, when the lightning is about 5 miles away).

The following material on lightning protection for land-based radio facilities was extracted from *Electrical Protection Guide for Land-Based Radio Facilities*, by David Bodle, copyright 1971 by Joslyn Electronic Systems, a division of Joslyn Mfg. and Supply Co. Inc. For a complete copy of the text, write to Gerald F. Croteau, Joslyn Electronic Systems, Santa Barbara Research Park, P. O. Box 817, Goleta, California 93017.

## GENERAL

*Scope*

Land-based mobile and microwave radio stations operate in a hostile electrical environment because of their exceptionally severe exposure to lightning. A variety of protection measures are required to assure uninterrupted operation. In addition to the primary objective of personnel safety, the following major parts of an installation require specific consideration:

Antennas and supporting structures
Coaxial lines and waveguides
Buildings and equipment enclosures
Radio, multiplex, carrier and switching equipment (where applicable)
Connecting facilities—communication and power land lines.

The following discussion will evaluate exposure conditions and give application details of the protection measures necessary to achieve an adequate protection level. The objective is to optimize the amount of protection relative to effective personnel protection, maintenance costs, and circuit outage time.

*Exposure Considerations*

The exposure of radio installations to lightning is greater than ordinarily experienced by other communication plants. Hill-top locations are sought to facilitate transmission, which not only substantially increases the incidence of strokes to the antenna structure but also presents conditions generally unfavorable for grounding. Protection requirements, not transmission, dictate the nature and extent of grounding at these sites. Fundamentally, it is preferable to "dump" as much stroke current directly to earth at the station as is economically feasible. This reduces the proportion of stroke current seeking remote earth over connecting land lines and, thus, simplifies the task of protecting these facilities.

Antennas and supporting structures are prime targets for lightning. Microwave "horns" and "dishes" are unlikely to be damaged by direct strokes. Mobile-radio antennas may be damaged to a limited extent by exceptionally heavy strokes through fusing of dipole elements at arc contact points.

*Engineering Aspects*

As a prerequisite to the development of effective remedial protection measures, it is necessary to define quantitatively the hazards involved and then establish specific protection objectives. Even in areas of low thunderstorm incidence, the possibility of a direct stroke to the antenna structure is strong enough to require protection against the hazard of electric shock to operating personnel. However, there is a lower probability of plant damage, so there is enough latitude to exercise economic discretion with regard to protection requirements. Probability figures concerning stroke incidence per thunderstorm day can be developed quite easily for specific plant arrangements and locations. This method provides a practical engineering basis for optimizing the amount of protection employed, and it is preferable to a brute force approach, which is likely to lead to gross overprotection.

*Evaluation of Lightning Exposure*

Thunderstorms are generally of two types: (1) convection storms, which are local in extent and of relatively short duration; and (2) frontal storms, which extend over greater areas and may continue for several hours. Storms of the convection type account for the majority of annual thunderstorm days in North America, yet experience indicates that they produce less plant damage than thunderstorms of the frontal type.

Frontal thunderstorms result from the meeting of a warm, moist front and a cold front; this interface may extend for several hundred miles and exposes large areas to particularly severe and destructive lightning discharges. Conditions in the southeastern part of the United States and in some of the midwestern states are particularly conducive to frontal storms. They tend to predominate in the spring and early summer but are also occasionally experienced during the winter (see Figure 4-9). Such midwinter storms can be particularly destructive when they occur in conjunction with snow.

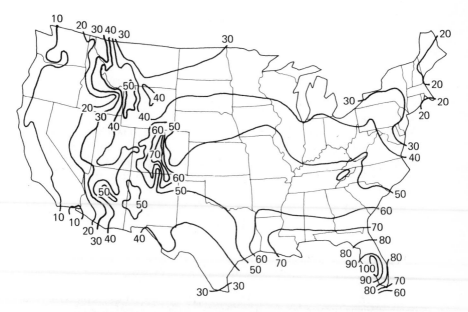

**Figure 4-9.** Mean annual number of days with thunderstorms in the United States. (Courtesy of Joslyn Electronic Systems, a division of Joslyn Mfg. and Supply Co., Inc.)

Because the antenna tower of a communications facility is the prime lightning target in a radio installation, the possibility of a direct stroke to the equipment building is remote (see Figure 4-10). However, the connecting communication and power land lines require attention because of their exposure to the high earth potential gradient that develops in the vicinity of a radio station at the time of a stroke to the antenna. In addition, these land-line facilities are still subject to their normal exposure outside of the station area.

Together these sources of exposure place the connecting facilities in a vulnerable position. The degree of protection justified for these facilities will depend on how essential they are to the operation of the station. In addition to the problem of damage to connecting facilities, it is essential to note that surges originating on a power line serving a station are frequently of sufficient magnitude to cause trouble in station apparatus.

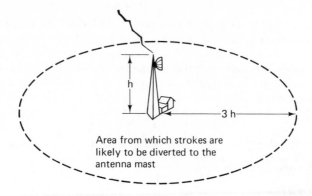

h

——— 3 h ———

Area from which strokes are
likely to be diverted to the
antenna mast

**Figure 4-10.** Strokes to a structure on level terrain—area of influence of mast. (Courtesy of Joslyn Electronic Systems, a division of Joslyn Mfg. and Supply Co., Inc.)

Arresters used on the primaries of distribution transformers have sparkover and discharge voltages too high to protect equipment on utilization circuits such as are used in radio stations. Secondary arresters having much lower sparkover and discharge voltages are required to provide adequate protection.

## FUNDAMENTAL PROTECTION MEASURES

*Introduction to Grounding*

Grounding involves more than providing a connection between an electrical circuit and some form of electrode in contact with the soil. It can be an effective protection measure, but to design a practical grounding system that will fulfill all requirements of a particular situation is a complex task. The reasons for grounding are to:

Reduce the hazards of electric shock

Protect wiring and circuit components in limiting extraneous overvoltages

Facilitate rapid deenergization of faulted power circuits

Provide a nondestructive path to ground for lightning currents contacting structures

Provide paths to ground for longitudinal (common mode) shielding currents in metallic cable shields, thus reducing induced currents in cable conductors.

*Grounding Philosophy*

If a conducting path for lightning stroke currents is provided between the point of contact of a stroke to a structure and a suitable grounding electrode, physical damage and shock hazards may be substantially reduced.

However, as will be discussed later, supplemental measures are required, in addition to simple grounding, to obtain highly effective protection.

Grounding can reduce electric shock hazards in two ways. First, it can limit potentials to ground of conducting objects within areas occupied by personnel. Second, it can shorten the time of possible exposure to excessive potentials by facilitating the rapid operation of fault current interrupting devices. The need for supplemental bonding between conducting objects in order to equalize potentials in the presence of steep wavefront surges experienced with lightning is discussed below.

### Bonding for Equalization of Potentials

Grounding is an important element of most protection arrangements, but it is only one of several measures necessary to achieve an effective level of protection. Unfortunately, the notion still exists that all protection problems can be solved by simply providing a connection to earth. This, of course, is simply not true. *All grounding electrodes have a finite resistance to earth*, even such extensive structures as metallic water pipes, *and to this must be added the impedance of grounding conductors.* Obviously, more must be accomplished than can be secured by simple grounding. This suggests potential equalization, which is obtained by frequent bonding (interconnection) of all conducting components of an installation (see Figure 4-11). In addition, supplemental conducting paths should be provided to

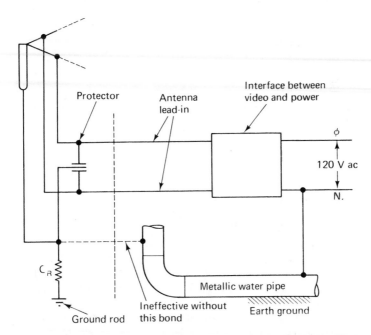

**Figure 4-11.** Protection of TV receiver. (Courtesy of Joslyn Electronic Systems, a division of Joslyn Mfg. and Supply Co., Inc.)

a ground electrode system to reduce the impedance of grounding paths to a practical minimum.

With bonding it is possible to approach an equipotential zone throughout an installation, thus assuring personnel safety and also contributing to equipment protection. Figure 4-12 illustrates the weakness of simple grounding arrangements and the need for supplemental bonding. Only by placing a bonding conductor between the cabinets will dangerous voltages in the operating area be eliminated.

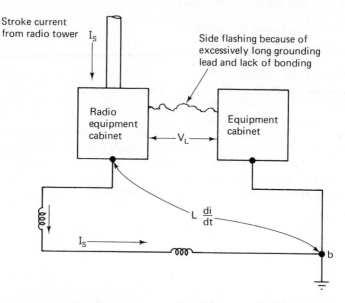

**Figure 4-12.** The weakness of simple grounding arrangements. (Courtesy of Joslyn Electronic Systems, a division of Joslyn Mfg. and Supply Co., Inc.)

*Voltage Limitation*

Protectors and arresters are usually open-circuit devices that pass no significant current at the normal operating potentials of the circuits to which they are connected. However, they are capable of sparking over on extraneous voltages at predetermined values and discharge current, usually from an energized conductor, to ground. During the period of the discharge, such a device will limit voltage across its terminals to values sometimes less, but rarely greater, than the initial sparkover value.

*Protectors* are effective devices for limiting voltages on communication circuits that can tolerate the shunt capacitance they introduce (a fraction of a picofarad). Protectors should be distinguished from *arresters*. Arresters are designed for use on power circuits. They are larger in size than protectors to permit the discharge of higher current, but, more importantly, they interrupt the flow of steady-state current, normally on the circuit, promptly after discharge of the surge or transient current is completed. The use of protectors is restricted to communication circuits which typically operate at relatively low voltages and currents.

Communication protectors typically consist of gas tube devices or closely spaced, flat carbon electrodes (carbon blocks) discharging in air. The gas tubes use metal electrodes in an enclosure with inert gas at reduced pressure. The gas-tube protector provides essentially the same level of protection as the carbon block device, but the wider gap spacing gives a much longer service life with a commensurate reduction in maintenance.

### Current Interruption

Fuses and circuit breakers are not satisfactory for interrupting lightning surge current because of inherent time delay. Such surges must be diverted to other paths having adequate current-carrying capability. For example, an excessive and destructive current can be bypassed around a vulnerable component by means of a discharge gap, selected so that it will spark over before the destructive current goes through the component.

## ANTENNAS AND SUPPORTING STRUCTURES

### General

Metal antenna structures are inherently self-protected. Some mobile radio antennas and most antenna horns, dishes, and reflectors are unlikely to be damaged by direct lightning strokes, but they should be bonded to metal supporting structures to eliminate subsequent arcing. If the ability of an antenna to withstand direct lightning strokes is doubtful, an air terminal to intercept strokes should be provided where the transmission pattern permits. Air terminals may be attached directly to a metallic supporting structure and should protrude sufficiently above vulnerable elements to provide an adequate *cone of protection*. Top lighting fixtures may also be subject to damage if they are not shielded either by an antenna or an air terminal.

Metallic antenna towers, either guyed or self-supporting, provide an excellent conducting path for stroke currents, but the footings, base, and guy anchors of such structures must be properly connected to suitable grounding electrodes. [According to J. L. Marshall in *Lightning Protection*, (John Wiley & Sons, N.Y., 1973), guy wire insulators near the tower of a VHF antenna should be bypassed with protective spark gaps for lightning protection.]

When wood poles are used to support antennas or passive reflectors, an air terminal should be provided at the top of the pole to intercept strokes. This will give protection against pole splitting and possible antenna damage. In a common arrangement, a ground rod is attached to the pole with one end protruding sufficiently above the top of the pole to provide a suitable cone of protection. A #6 AWG, bare-copper, down lead is connected to the rod and stapled directly to the pole on the side opposite the line or waveguide to the antenna. All pole-top hardware, the antenna, and any supporting guys should be bonded to this grounding conductor. At the base of the pole the shields of lines or waveguides, equipment cabinets, and any other conducting objects should be bonded to the down lead, which then must be connected to the common area ground system.

### Microwave Antennas

Paraboloid (or *dish*) antennas and horn reflector antennas commonly used for microwave transmission are usually rugged enough to sustain direct lightning strokes without significant damage.

### Grounding of Antenna Support Structures

*Mobile-radio Antenna Support Structures.* A grounding arrangement typical of a wood pole-mounted installation is shown in Figures 4-13 and 4-14. The antenna selected for this illustration is a stacked, folded dipole which may be top (as shown) or side-mounted. The support mast is heavy-wall duralumin topped by a cast aluminum cap high enough above the top dipole element so it can intercept lightning currents. The cap and mast are capable of conducting lightning currents. Therefore, a suitable path to ground is provided by a down lead from the mast to the base of the wood pole, where it is connected to grounding electrodes (Figure 4-13).

If mobile-radio antennas are installed on top of buildings, the cable shields, down conductors, and any other conducting objects within six feet of the base of the antenna-supporting structure should be commonly bonded. Then the entire arrangement should be connected by separate conductors to two grounding media within the building, such as metallic water pipes or building steel, or by two separate conductors to a suitably "made ground."

Where antennas are supported on metal towers, grounding is simplified. In such cases, antenna masts, cable shields, and all pole-top hardware are connected to the tower, which now serves as a grounding path for lightning currents. Suitable connections from the base of the tower to an adequate ground electrode arrangement are, of course, required.

*Microwave Antenna Support Structures.* Since microwave antennas and their metal supporting structures are essentially self-protected, very little in the way of special protection is required at the top of the tower. However, adequate grounding of metal supporting towers requires consideration. Grounding arrangements at a microwave station are not dictated by radio operation requirements but rather by protection considerations. A low impedance connection to earth is not necessarily required. If potential equalization is obtained throughout the station area by effective bonding, the hazard to personnel and station equipment will be minimal. However, some proportion of antenna stroke current should be dissipated directly into the earth at the station site. This diverts excessive current from the

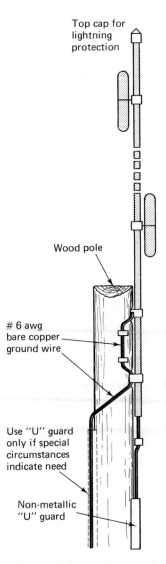

**Figure 4-13.** Grounding details for a pole-mounted antenna. (Courtesy of Joslyn Electronic Systems, a division of Joslyn Mfg. and Supply Co., Inc.)

connecting land facilities and thus reduces damage to such facilities. It is good practice to "dump" to earth through a station grounding structure as large a proportion of lightning stroke current from the antenna tower as is economically feasible.

There is an optimum balance between the cost of constructing a low-resistance ground at a station and the cost of protecting connecting com-

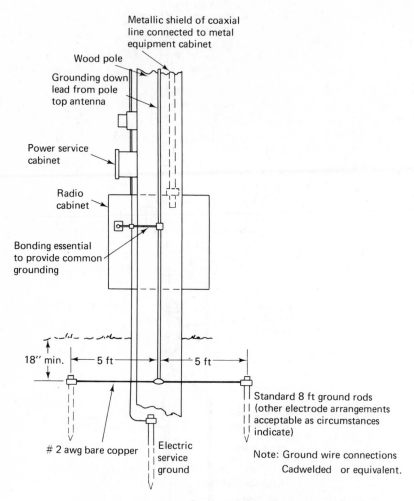

**Figure 4-14.** Representative grounding arrangement of a pole-mounted installation. (Courtesy of Joslyn Electronic Systems, a division of Joslyn Mfg. and Supply Co., Inc.)

munication and power land lines so there will be no damage or outage due to a high-surge current. It is more practical to specify in engineering instructions to field personnel only the physical configuration and dimensions of a station grounding arrangement than to insist on a specific ground electrode resistance.

Where local grounding conditions are very poor or exposure is especially severe, additional protection for communication and power lines should be provided. Economics may also suggest the advisability of lowering station ground impedance in conjunction with such a program. The mechan-

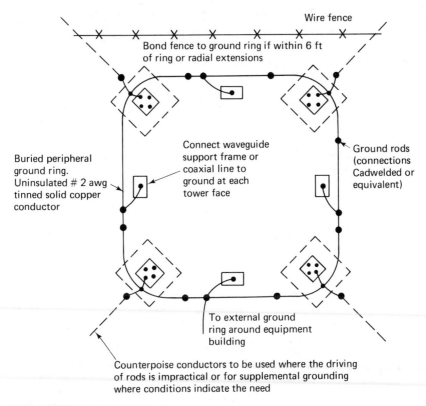

**Figure 4-15.** Grounding details for a free-standing metal tower. (Courtesy of Joslyn Electronic Systems, a division of Joslyn Mfg. and Supply Co., Inc.)

ics of implementing these protection objectives will be discussed further. However, some details are shown in Figures 4-15 through 4-18.

In recent years, considerable attention has been directed to the effectiveness of concrete-encased grounding electrodes. However, in practice there is no guarantee that the hardware and reinforcing in tower footings and guy anchors are electrically continuous. Consequently, it seems advisable to continue the established custom of providing supplemental grounding at such locations as added assurance against spurious arcing. Also, the perimeter ground ring shown around tower footings affords, in addition to local grounding, a convenient means of interconnecting the tower to the buried ground ring of the equipment building, as well as supplemental counterpoise wires, should circumstances require it.

*Antenna Support Structures on Buildings.* Antenna towers are only occasionally mounted on buildings in rural areas; in urban areas, however, it is common practice. In the latter case, the buildings are frequently large

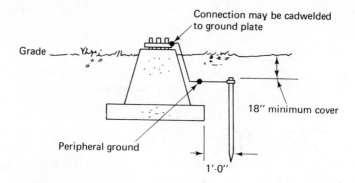

**Figure 4-16.** Grounding arrangement at tower pier. (Courtesy of Joslyn Electronic Systems, a division of Joslyn Mfg. and Supply Co., Inc.)

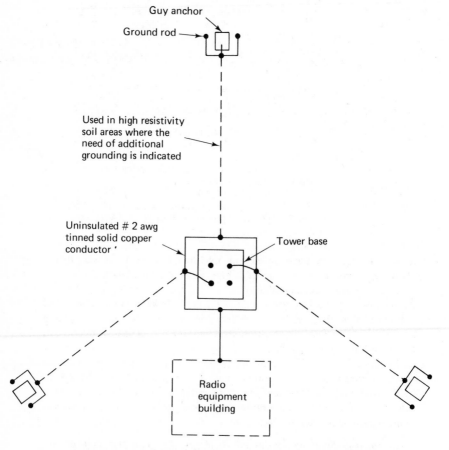

**Figure 4-17.** Grounding arrangement for a guyed tower. (Courtesy of Joslyn Electronic Systems, a division of Joslyn Mfg. and Supply Co., Inc.)

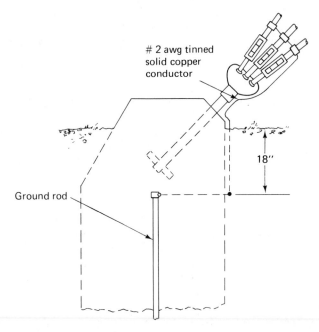

**Figure 4-18.** Grounding details for an antenna guy anchor. (Courtesy of Joslyn Electronic Systems, a division of Joslyn Mfg. and Supply Co., Inc.)

steel frame structures in which the radio equipment only occupies a small part of the available space. In such cases, the structural steel offers an excellent ground path for roof-mounted antenna towers and associated radio equipment. Metal antenna support towers should be bonded to building steel with #2 AWG copper conductors by two separate paths. In addition, it is good practice to bond the tower at roof level to a large metal water pipe. Frequently, pipes feeding water tanks on the roof are available and are suitable for grounding purposes. As a precaution, it should be established that the building structural steel is adequately interconnected with the water mains and power grounding electrode.

Guy wires associated with an antenna tower mounted on the roof of a building should be grounded at their base in essentially the same manner as has been described for grounding towers. However, one ground connection per guy should be adequate. Also, if there is less than a 6-foot clearance between guy wires, support towers, grounding and bonding conductors, cable shields, etc. and any other conducting object, a bonding connection using a #6 AWG copper conductor should be placed between them to prevent the possibility of side flashing.

Waveguides and the shields of coaxial lines should be bonded to the base of the tower before the lines enter the radio equipment area. The general practices described below for radio equipment buildings are also applicable to cases where the radio equipment occupies only a small section of a larger building. Fundamentally, the equipment should be well bonded to equalize potentials and then grounded as a unit to the same grounding system employed for the antenna tower; e.g., structural steel frame, water pipes, protection down leads.

## COAXIAL LINES AND WAVEGUIDES

Coaxial lines are subject to two possible hazards from surge currents flowing in the outer conductors. One hazard is the induced surge voltages that may damage line dielectric or associated equipment, and the other is the mechanical crushing forces. The solution to both problems is relatively simple—merely provide a shunt path to ground for antenna stroke currents. This is easily accomplished in the case of metal poles and towers by bonding the line to the conducting structure at the top and bottom. Where lines are supported at intermediate points, supplemental bonding at such points will eliminate possible arcing. In the case of wood supporting structures, supplemental conduction can be satisfactorily provided by a parallel conductor as described for antenna support structures in the previous section.

Waveguides supported on metal structures should also be bonded to the structure at the top and bottom and also at intermediate points of support. This will provide sufficient additional conductivity in shunt with flexible sections to prevent possible damage at these critical points. Where the supporting structures cannot be used for this purpose, conducting bonds should be placed across all flexible waveguide sections.

## PROTECTION OF STATION BUILDING, PERSONNEL, AND EQUIPMENT

*General*

A station building is not likely to be struck directly by lightning because of the shielding provided by the antenna support structure. However, waveguides and the shields of coaxial lines can conduct hazardous currents from the antenna into the building unless protective measures are employed. A station grounding system is essential to divert a large proportion of the stroke current directly to ground before it enters the building. The grounding system must also be designed to reduce earth potential gradients significantly under and around the station building by employing a distributed electrode configuration surrounding the building. The most frequently used arrangement for achieving these objectives is a buried conductor around the outside perimeter of the building, supplemented by ground rods. The "ring ground" arrangement commonly used consists basically of a bare conductor large enough to withstand physical hazards, placed about 2 feet from the outside of the foundation and buried at a depth of 18 to 24 inches. This buried counterpoise wire is supplemented by ground rods when conditions permit their installation.

Engineering considerations dictate that the common grounding system at a radio site should conduct to earth as large a proportion of the antenna stroke current as is economically feasible. The object is to optimize the balance between the cost of constructing a low-resistance ground at the station site and the cost of supplemental protection for the connecting facilities. The resulting protection should be sufficient to conduct remnant stroke current safely to remote points without damaging or interrupting the functioning of the facilities.

### Design of Station Grounding Electrode Systems

The grounding system selected must, in addition to optimizing resistance to earth, employ a configuration that minimizes earth potential gradients under the station building and throughout the immediate area of the building. It is desirable that a ground be effective throughout the year and not deteriorate significantly during cold weather. Although the incidence of lightning is low during the winter months, severe lightning strokes to earth do occur often enough in winter, during unsettled weather conditions, to warrant consideration. Freezing of the earth around an electrode will significantly increase its resistance to remote earth and thus substantially degrade its grounding effectiveness.

Where frost penetrates below the depth of the counterpoise conductor, the ring grounding arrangement will suffer a substantial loss in effectiveness. The only part of such a structure not affected will be some portion of the rods. In areas experiencing deep frost penetration in the earth, the ground electrode system should be placed deep enough to avoid exposure. This can be accomplished by installing the ring wire and rods at or near the bottom of the trench excavated for the building footings. An alternate solution is a concrete-encased electrode.

### Bonding and Grounding Within a Station

The limitation of inductive voltages within a station operating area is an essential personnel safety measure. Judicious bonding is also required to enable protective devices to achieve their full effectiveness.

Experience has established the desirability of providing a *grounding bus* around the inside of equipment buildings to facilitate grounding and bonding of equipment enclosures and metallic objects therein. One common form of bus is a #2 AWG copper conductor attached directly to the wall. One ring bus of this type is usually adequate for moderate size buildings without partitions. In larger structures, the use of lateral runs attached to partitions is advisable. Each end of supplemental bus runs should be connected to the main perimeter bus.

The internal ground bus must be effectively connected to the external grounding ring. Several interconnections should be provided, the exact number depending on the dimensions of the building. Incidental objects within a building should also be bonded to the internal ground bus. These would include such items as intake louvers, filter frames, metal doors, and frames.

It cannot be overemphasized that bonding, despite its simplicity, is an effective protection measure for the reduction of hazardous surge voltages.

### Materials for the Construction of Grounds

Bare copper conductors and copper-clad ground rods have been used extensively for the construction of grounds and are still in common use. The durability of these materials has been proven by experience.

### Size of Grounding and Bonding Conductors

When installed indoors and suitably protected from physical damage, #6 AWG copper wire is electrically adequate for bonding and grounding. However, experience has shown that for direct burial in the earth, a #2 AWG solid-copper conductor is preferable because of possible exposure to mechanical stress or damage. Also, under these conditions a solid conductor should be used, since it is less subject to corrosion than a stranded conductor of similar gauge. Inside a building, the latter may safely be used for grounding and bonding.

### Method of Construction

Some form of metal fusing, such as Cadwelding, provides an effective and durable means of connecting the components of a grounding system. Pressure connectors sometimes loosen due to cold flow of the conductors.

## PROTECTION OF LAND COMMUNICATION FACILITIES

### General

Land communication facilities in metropolitan and other built-up areas are usually shielded from lightning. The shields of communication cables in such areas are well grounded, and extensive metallic piping systems mitigate the development of hazardous earth potential gradients. A radio antenna connected to such an extensive ground electrode system should not significantly change the local exposure conditions. Therefore, land communication facilities should not require any more protection than that normally provided.

Cables and lines associated with radio installations outside of built-up areas are usually exposed to lightning. Grounding conditions are likely to be poor, requiring special protection to prevent serious damage to connecting facilities.

### Protection of Wire Lines and Associated Facilities

Open wire plant has substantially higher insulation strength than paired telephone cable and station apparatus. The likelihood of lightning damaging open wire lines is relatively small, and they can support high voltage surges that damage weaker connected plant, such as paper-insulated cable. Consequently, protection should be provided wherever exposed open wire lines interface with other types of communication plant.

Wire plant serving a radio station (see Figure 4-19) should be terminated on the outside wall of the building, permitting the ground lead from the protectors to be run directly to the external buried ground ring. The duty on these protectors will be severe, making it advisable to use gas-tube protection to reduce the possibility of permanent grounding. Gas-tube protection may be applied in either of two ways:

1. Combination of fail-safe gas tubes in a fuse-type mounting listed by UL for this purpose.

2. Gas tubes in parallel with the carbon blocks in a UL-listed fuse-type mounting.

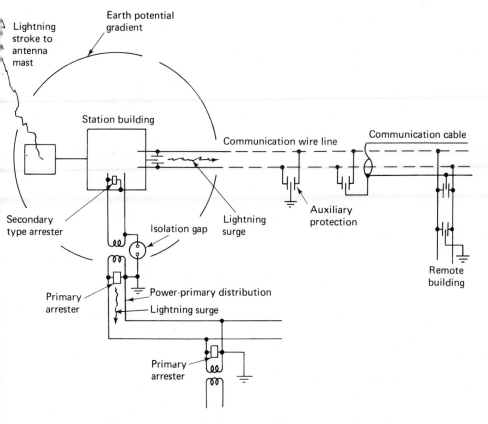

**Figure 4-19.** Exposure of connecting facilities as a result of a stroke to the antenna mast. (Courtesy of Joslyn Electronic Systems, a division of Joslyn Mfg. and Supply Co., Inc.)

At the junction of wire plant and Plastic Insulated Cable (PIC), protectors should be applied to all wires entering the cable. This is another point where the use of gas tubes rather than carbon blocks is preferable because of their greater freedom from permanent grounding. PIC, rather than paper-insulated cable, should be used for service to radio stations because of its superior dielectric strength. Unfortunately paper-insulated plant may be the only type available at some locations. Since paper-insulated cable pairs are especially vulnerable, it is good practice at open-wire junctions to place protection on all cable pairs even though they will usually outnumber the open-wire pairs. Also, if there are drops working out of the cable in the vicinity of the junction, protected terminals should be used at such points. This practice provides more effective grounding of the cable shield. Another means of reducing the magnitude of surges on an open wire line is through the use of auxiliary protection. This entails the installation of protectors between each line wire and a ground electrode constructed at some point within one-quarter to one-half mile of the junction where the soil conditions appear to favor grounding. In all such special protection measures, gas-tube protectors, selected according to the application, will provide effective and relatively trouble-free service.

Cables serving radio stations outside of built-up areas are usually subjected to large potential differences between the station terminal end and points remote from the station site. This is especially true with stations situated on hills, since the soil tends to be dry and rocky, presenting unfavorable conditions for grounding. Frequently, a mile or so away at a lower elevation, the soil will have a lower resistivity, thus offering an attractive "sink" for surge currents resulting from strokes to the antenna.

For a radio station large enough to require more than just a few circuits, a metallic sheath cable is generally used, usually a small-diameter cable branching from a larger feeder cable that serves other customers. The shield resistance of a small-diameter cable is high. Consequently, large voltages can be produced between the core conductors and the shield by surge currents of even moderate magnitudes flowing in the shield. Where the severe nature of this exposure is not recognized, and the protection approach has been rather casual, major damage has resulted to the cable plant and associated facilities. Because of the severity of this exposure, *PIC should be used for service to radio stations.*

*Aerial Cable Plant.* Station-type protection is required for aerial cable plant application. Because of the severe duty on protectors, a gas-tube device will reduce maintenance and circuit outages. Two approved methods may be used:

1. Gas tubes in parallel with the carbon blocks in UL-listed fuseless house terminal or station mounting.
2. The combination of a fail-safe gas tube in a fuseless terminal or mounting listed by UL for the purpose.

Experience has established the desirability of the following cable protection measures:

All pairs in the branch cable should be brought into the station.

Protectors should be provided at the station terminal on all working pairs in the cable. The nonworking pairs should be connected to the ground lug in the terminal. This arrangement equalizes potentials in the cable, and the grounded nonworking pairs provide shielding for the working pairs.

The cable shield should be bonded to the terminal, and then both should be grounded by connection to the internal station ground bus. This ground connection must be run in a straight line to minimize its length.

When a cable enters a station aerially, the cable support strand should be bonded to the station ground.

In the case of an underground entrance using metallic conduit, the cable shield should be bonded to the conduit at each end of the run to prevent arcing. Also, bonds should be provided between the aerial strand and the conduit, and between the conduit and the cable terminal. If non-

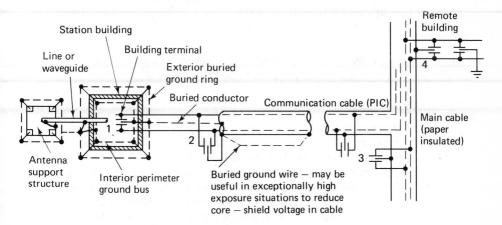

Notes:
(a) If cable is run into station building, bond shield to external
    ground ring at entrance — protection at point 2 is not required.

(b) If main cable is PIC insulated, protection at point 3 is not required —
    protection at point 4 may still be useful to provide a path to ground
    for surge current on cable shield.

**Figure 4-20.** Protection of connecting aerial cable facilities. (Courtesy of Joslyn Electronic Systems, a division of Joslyn Mfg. and Supply Co., Inc.)

metallic conduit is used, #6 AWG copper conductor should be run between the aerial strand and a grounding lug in the terminal, instead of bonding to the conduit. This assures electrical continuity between the strand and station ground.

Where an aerial branch cable enters a main feeder cable, protectors having a sparkover voltage less than about 1,000 peak volts should be provided on all pairs in the branch cable (see Figure 4-20). Also, protectors should be placed on drops working out of the main cable within 1,000 to 1,500 feet from the junction. If the main cable has paper insulation, it will be susceptible to dielectric failure from the higher voltage surges propagated in the PIC branch cable. If this problem should develop in the main cable, two special protection measures may be used. First, apply protectors at the junction to all pairs in the main cable not previously protected; second, install a buried counterpoise wire starting at the station end and running along the right-of-way under the cable for about 1,000 feet. This counterpoise should be connected to the station ground and will divert substantial current from the shield of the branch cable.

*Buried Cable.* In rural, high-exposure areas, aerial construction is most commonly used. However, if buried cable is to be used, PIC having an extruded core insulation is preferable. This is a high dielectric cable referred to as PAP or PASP and is especially suited for applications involving high lightning exposure.

## PROTECTION OF POWER SERVICES
## AND UTILIZATION EQUIPMENT

### Need for Protection

Experience has shown that ac-power radio equipment and tower lighting components are generally vulnerable to damage from lightning surges and abnormal transients on the power distribution facilities.

*The burden' of providing adequate protection to assure service continuity and minimize maintenance expense must be assumed by the user.* It is definitely the responsibility of the users of utilization equipment to protect it against hazards unique to a specific application. In the case of radio stations, protection against the following hazards should be considered.

1. Interchange of surge current between station power equipment and a power line serving the station when antenna supporting structures are struck by lightning—earth potential gradients associated with such strokes may be great.
2. Exposure of station equipment (including tower lighting facilities) to surges produced by lightning striking the primary power distribution line outside of the cone of protection of the antenna tower.

*Protection Measures*

An *arrester* is a discharge device used on power circuits to limit abnormal surge and transient potentials. In this text we are only concerned with arresters for low voltage circuits (600 volts and less), which are frequently referred to as *secondary type arresters*. The term used to identify the simple discharge gap used on communication circuits is *protector*. Although both

**Figure 4-21.** Joslyn Model 2001-83 spark gap. (Courtesy of Joslyn Electronic Systems.)

**Figure 4-22.** Joslyn Model 1920-02 arrester, used on loran antenna systems. (Courtesy of Joslyn Electronic Systems.)

**Figure 4-23.** Joslyn Model 1245-01, a single-phase lightning arrester for a radio station. (Courtesy of Joslyn Electronic Systems.)

these devices discharge current resulting from abnormal voltages, they are quite different in construction and application.

Arresters are used on circuits having appreciable, steady-state voltage; therefore, a simple discharge gap would continue to conduct steady-state current after the abnormal surge that initially operated the gap has attenuated. The steady-state current flowing through an arrester after the surge has attenuated is referred to as *power-follow current* and cannot be tolerated for any appreciable time because: 1. current disconnect devices will operate and deenergize the circuit; and 2. the arrester will be damaged. It is necessary, therefore, that arresters incorporate some "clearing" mechanism in addition to the discharge gap to promptly interrupt the flow of power-follow current (see Figures 4-21 through 4-23.)

The arresters that a power utility installs on its primary distribution circuits adjacent to a transformer are intended exclusively to protect the transformer. They do not provide for secondary circuits and the associated utilization equipment. Figure 4-24 shows an arrester on a single phase distribution circuit of a multigrounded neutral (MGN) system.

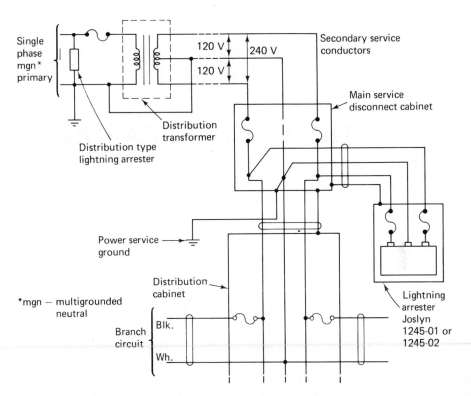

**Figure 4-24.** Typical power service for a small-to-medium size radio station. A single lightning arrester provides protection for all utilization equipment. (Courtesy of Joslyn Electronic Systems, a division of Joslyn Mfg. and Supply Co., Inc.)

## CATV SYSTEMS

Field experience has shown that many of the head-end receiving stations of CATV systems are not adequately protected. As a result, lightning damage occurs at antenna sites, and from these points, surge voltages travel over the cable facilities and damage line amplifiers. Receiving antennas are highly exposed structures, and to assure service continuity and reduce maintenance expense, the same degree of protection should be provided as is employed at the transmitters. This involves adequate grounding and bonding, protection of cable facilities, and secondary arresters on the commercial power facilities to the station.

# 5

# RADIATION HAZARDS

## INTRODUCTION

Since 1940 the number of radar systems, television and broadcast transmitters, communications systems, X-ray machines, and lasers has increased at a fantastic rate. As shown in Tables 5-1 through 5-3, there are many thousands of these electromagnetic radiators now in use. And not only are these devices being manufactured at greater rates than before, they are being operated at much higher levels of radiation. The radiation from these sources is permeating the modern environment, so that each of us is exposed to it daily, thus creating still another environmental hazard.

Because of this permeation, there is growing concern, even anxiety, among some health physicists that even relatively low power densities from these transmitters may be polluting our bodies. There is no question that high-level radiation can kill or blind you. But the effects of low-level radiation are not so obvious. Long term exposure may create undreamed-of biological changes. Our knowledge of such effects is rather skimpy and is based more on circumstantial evidence than on cause-effect proofs. Nevertheless, as long as low-level radiation is suspect, you should exercise caution around any transmitter. To wait for absolute proof of a danger may cause you unnecessary suffering or possible disaster.

The problem with electromagnetic radiation—whether it comes from a chest X-ray machine or a military search radar—is that it can be both mysterious and insidious. Usually radiation cannot be felt, seen, heard, tasted, or smelled, although lasers, of course, can be seen, and microwaves at certain frequencies may create a feeling of warmth.

TABLE 5-1  *Sources of ionizing radiation*[1]

| Product | Estimated Inventory Jan. 70 |
|---|---|
| Color TV (households) | 24,000,000 |
| Medical X-ray | 115,000 |
| Dental X-ray | 100,000 |
| Industrial X-ray | 15,000 |
| Accelerators | 1,200 |
| Electron Microscopes | 500 |

[1]Jesse Y. Harris, "National Inventory of Electronic Products," *Electronic Product Radiation and the Health Physicist*, BRH/DEP 70–26 Bureau of Radiological Health, Dept. of Health, Education, and Welfare (October 1970).

TABLE 5-2  *Sources of microwave radiation*[1]

| Product | Estimated Inventory Jan. 70 |
|---|---|
| Communications Transmitters | 66,000 |
| Domestic Ovens | 50,000 |
| Commercial Ovens | 45,000 |
| Diathermy Units | 15,000 |
| Radar, Pleasure Boats | 7,500 |
| Radar, Stationary | 5,500 |
| Industrial Heating | 300 |

[1]Jesse Y. Harris, "National Inventory of Electronic Products," *Electronic Product Radiation and the Health Physicist*, BRH/DEP 70–26 Bureau of Radiological Health, Dept. of Health, Education, and Welfare (October 1970).

TABLE 5-3  *Inventory of radio and lf devices, Jan. 1970*[1]

| Device | Estimated No. |
|---|---|
| Transmitters | |
| Commercial AM | 4,300 |
| Commercial FM | 2,200 |
| Educational FM | 400 |
| Commercial TV | 700 |
| Educational TV | 200 |
| Amateur | 300,000 |
| Citizens | 3,000,000 |
| Aviation Services | 200,000 |
| Industrial Services | 1,700,000 |
| Transportation Services | 500,000 |
| Marine Services | 200,000 |
| Public Safety Services | 700,000 |
| TV Translators and Boosters | 2,200 |
| Diathermy Units | Several Thousand |

[1]Jesse Y. Harris, "National Inventory of Electronic Products," *Electronic Product Radiation and the Health Physicist*, BRH/DEP 70–26 Bureau of Radiological Health, Dept. of Health, Education, and Welfare (October 1970).

To make matters worse, there may be a long wait between radiation exposure and the resulting injury. By the time you develop the symptoms, you may have forgotten the cause. Not knowing about the dangers to which you are exposed may leave you defenseless.

To place radiation problems in perspective, however, it should be noted that the average man on the street has little to worry about as far as radiation is concerned. On the other hand, the electronics technician or engineer may have his face and hands in, around, and about transmitting devices all day long, perhaps during every working day of the year. In such positions he may have shields removed, interlocks bypassed, and controls set at maximum, thereby setting himself up for maximum exposure to radiation. Still worse, he may be working with exotic futuristic devices whose hazards are barely imagined, much less documented and brought to his attention. For this worker, radiation may be an uncomfortable fact of life, a work-related hazard; certainly he should be made aware of the dangers involved and the safety precautions that he can take.

This chapter discusses some of the more flagrant sources of danger in color TV, radar, microwave ovens, klystrons, and cold-cathode gas discharge tubes; it also shows how this radiation is harmful to people and to electro-explosive devices. The chapter does not cover hazards from medical or industrial X-rays or nuclear devices, as many other texts provide such information. Hazards from lasers are so extensive that they are discussed in a separate chapter (Chapter 6).

Before considering the types of radiation and their biological effects, let's look at the government's role in controlling radiation hazards.

The Radiation Control for Health and Safety Act was enacted in 1968 to protect the public from unnecessary exposure to radiation from electronic products. To accomplish this, the Bureau of Radiological Health in the Department of Health, Education, and Welfare has been given the responsibility for developing and administering performance standards for controlling such radiation. The Bureau has published numerous reports describing radiation problems along with recommended solutions and has issued performance standards for television receivers, microwave ovens, and cold-cathode gas discharge tubes. (These standards have been compiled in *Regulations for the Administration and Enforcement of the Radiation Control for Health and Safety Act of 1968*, November 1972, which is available from the Government Printing Office for $0.45. Ask for stock number 1715-0043.)

Quasi-official bodies such as the National Council on Radiation Protection and Measurements and the American National Standards Institute are also involved in radiation control programs.

No matter how many agencies and groups are involved in the preparation of radiation protection limits and laws, it is still up to the individual technician to read and take heed of the regulations, the recommendations, the suggestions, and the laws. The government cannot protect you from acts of carelessness, ignorance, or recklessness.

## BIOLOGICAL EFFECTS

Radiation may be defined as energy moving through space either as electromagnetic waves or as invisible particles such as neutrons and beta particles. Here we shall consider only electromagnetic radiation. This radiation may be divided into two major types: ionizing and nonionizing. Ionizing

**TABLE 5-4** *Summary of maximum recommended levels of radiation for human exposure*[1]

| Country and Source | Radiation Frequency | Maximum Recommended Level | Condition or Remarks |
|---|---|---|---|
| U.S.A. (ANSI) | 10 MHz-100 GHz | 10 mW/cm² | Periods of 0.1 hr |
| | | 1 mW hr/cm² | Averaged over any 0.1-hr period |
| U.S. Army and Air Force | ... | 10 mW/cm² | Continuous exposure |
| | | 10 to 100 mW/cm² | Maximum exposure time in minutes at $W(mW/cm^2)$ $= 6000\ W^{-2}$ |
| | | 100 mW/cm² | No occupancy |
| Great Britain (Post Office Regulation) | 30 MHz-30 GHz | 10 mW/cm² | Continuous 8-hr exposure, average power density |
| NATO (1956) | ... | 0.5 mW/cm² | |
| Canada | 10 MHz-100 GHz | 1 mW hr/cm² | Averaged over any 0.1-hr period |
| | | 10 mW/cm² | Periods of 0.1 hr |
| Poland | 300 MHz | 10 μW/cm² | 8-hr exposure/day |
| | | 100 μW/cm² | 2 to 3 hr/day |
| | | 1 mW/cm² | 15 to 20 min/day |
| German Soc. Republic | ... | 10 mW/cm² | |
| U.S.S.R. | 0.1-1.5 MHz | 20 V/m | Alternating magnetic fields |
| | | 5 amp/m | |
| | 1.5-30 MHz | 20 V/m | |
| | 30-300 MHz | 5 V/m | |
| | 300 MHz | 10 μW/cm² | 6 hr/day |
| | | 100 μW/cm² | 2 hr/day |
| | | 1 mW/cm² | 15 min/day |
| Czech. Soc. Rep. | 0.01-300 MHz | 10 V/m | 8 hr/day |
| | 300 MHz | 25 μW/cm² | 8 hr/day, CW operation |
| | | 10 μW/cm² | 8 hr/day, pulsed |

[1]Richard A. Tell, "Broadcast Radiation: How Safe Is Safe?" © IEEE Spectrum (August 1972).

radiation can strip electrons from atoms and thereby create electrically charged ions that can disrupt life processes. Nonionizing radiation does not have the ability to create ions, but it can increase molecular vibration and rotation, thus creating heat. Either type can cause serious body harm, as described below (see Table 5-4 and Figures 5-1 through 5-4).

### Ionizing Radiation

The most common form of ionizing radiation is the X-ray. Because of its widespread use by physicians, dentists, and hospitals, it accounts for about 95% of the man-made radiation exposure each year in the United States. Without question, its benefits as a diagnostic tool in the healing arts far outweigh its dangers. Even so, most X-ray users recognize that X-rays must be treated with extreme caution. Very large doses may cause death or injury to the central nervous system immediately, or leukemia or cataracts later on. Through genetic damage, X-rays can cause birth defects in your children or in future generations.

There is no dose of X-radiation that may be considered "safe." No matter how low the level, any X-rays can cause possible damage to body cells. The damage may be temporary or permanent; the effects may not show up until years after the exposure. Generally, the doses are cumulative, with each exposure adding to the damage of the previous exposure.

The most common unit of measurement for ionizing radiation is the *rem*, which is a "measure of the dose of any ionizing radiation to body tissue

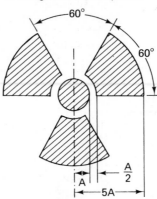

Radiation symbol

1. Cross-hatched area is to be magenta or purple.
2. Background is to be yellow.

**Figure 5-1.** X-ray radiation symbol. (Courtesy of The Federal Register.)

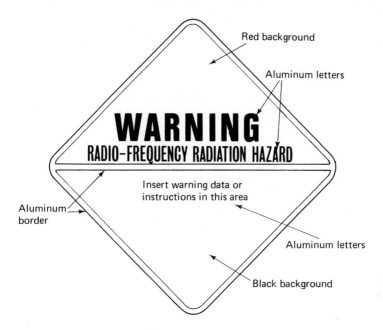

**Figure 5-2.** Radio-frequency radiation hazard warning symbol. (Courtesy of The Federal Register.)

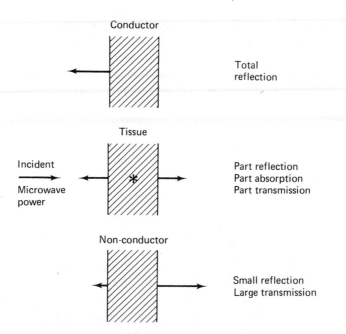

**Figure 5-3.** Interaction of microwave radiation with electrical conductors, biological tissue, and electrical insulators. (From Lawrence D. Sher, "Interaction of Microwave and RF Energy of Biological Material," Bureau of Radiological Health 70-26, Dept. of Health, Education, and Welfare.)

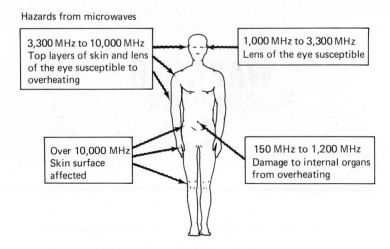

Hazards from microwaves

| 3,300 MHz to 10,000 MHz Top layers of skin and lens of the eye susceptible to overheating | 1,000 MHz to 3,300 MHz Lens of the eye susceptible |

| Over 10,000 MHz Skin surface affected | 150 MHz to 1,200 MHz Damage to internal organs from overheating |

**Figure 5-4.** Hazards from microwaves. (Reprinted from Electronics, September 30, 1968, Copyright McGraw-Hill, Inc., 1968.)

in terms of its estimated biological effect relative to a dose of 1 roentgen of X-rays."[1] For X-rays, one rem is equivalent to one *rad* which is the dose corresponding to the absorption of 100 ergs per gram of tissue.

The estimated total annual average whole-body dose from natural radiation (such as cosmic rays) is 102 mrem (thousandths of a rem). Miscellaneous radiation sources such as television and other consumer products add 2.6 mrem to this total. Medical radiation adds 72 mrem more, for an approximate total of 176 mrem per year.

Under the Occupational Safety and Health Act of 1970 the limits for the working man are:

Whole body: head and trunk; active
blood-forming organs; lens of eyes;
gonads  . . . . . . . . . . . . . . . . . . . . . . . . . . . . .$1\frac{1}{4}$ rem per calendar quarter
Hands and forearms, feet and ankles  . . . .$18\frac{3}{4}$ rem per calendar quarter
Skin of the whole body  . . . . . . . . . . . . . . .$7\frac{1}{2}$ rem per calendar quarter

An individual in a restricted area may receive doses to the whole body that are greater than these limits, provided that the dose to the whole body does not exceed 3 rems per calendar quarter and that the dose to the whole body, when added to the accumulated occupational dose to the whole body, does not exceed $5(N-18)$ rems where $N$ equals the individual's age in years at his last birthday. Any employee under 18 years of age may not receive a dose in excess of 10% of these limits in any period of one calendar quarter.

[1]From Federal Register, Vol. 36 No. 105, May 29, 1971.

For the electronics technician, the most common sources of X-rays are klystrons, magnetrons, thyratrons, and color TVs, all of which are discussed in later sections.

### Nonionizing Radiation

Nonionizing radiation is generated by radar, radio and television transmitters, and numerous other devices at power density levels that range from insignificantly low to dangerously high. Fences and warning signs may keep you away from the high-power radiations, but you can scarcely escape the low-level radiation unless you live in a copper screen room. At this time no one can say with certainty whether or not low-power radiation can cause you any bodily harm; it's just possible that there are subtle, long-term, cumulative effects from such radiation.

Both the frequency and the power level involved apparently play a significant part in any biological effect, but there is no general agreement on or definition of these frequencies and levels. Microwaves [300 megahertz (MHz) to 300 gigahertz (GHz)] cause the most concern, but frequencies as low as 100 hertz (Hz) are being studied for radiation hazards. Power levels from 10 microwatts per square centimeter to 100 milliwatts per square centimeter are under investigation.

Numerous experiments have been conducted with laboratory animals, but it is difficult to translate or extrapolate the results to human exposure limits. A biological effect, it should be noted, is not necessarily a biological hazard. Delayed effects of radiation make it necessary for careful observations to be made for several years before it can be said with certainty that particular radiations of particular frequencies and power levels cause specific damages. Fortunately, there have been no deaths due to radio or microwave radiation so far.

At the present time, the chief known effect of radio or microwave energy on living cells is to produce heat. If you are exposed to high intensity microwaves, you may feel warm, but don't count on it. At some frequencies this warmth may be felt too late to give you adequate warning of the danger.

The eyes and testes are particularly susceptible or vulnerable because of their poor ability to dissipate heat. In the case of the eyes, this heat can cause cataracts and other eye problems, as has been conclusively documented. Regular eye examinations have now become a way of life for technicians who work around radar and other microwave equipment. At one test site nearly 20% of the microwave workers have been found to possess the first indication of microwave eye damage. And note that the power level does not have to be high in order to cause damage; if you're close to the source, low power levels can create the same problem. In 1952 a technician developed a cataract as a result of working with a C-band magnetron with an average power output of only 250 watts.

The effect of microwave energy on the testes is to cause temporary or permanent sterility. It may also cause genetic damage. Researchers at Johns Hopkins University are studying an apparent coincidence of a high percentage of mongoloid children among radar workers and users.

Nonthermal effects—those that cannot be attributed to temperature elevations—under investigation include thyroid enlargement, headaches, loss of memory, change of pulse rate, genetic effects, and behavioral effects. This is only a partial list. The Bureau of Radiological Health is studying the possibility that some blood diseases may be caused by microwave radiation. Dr. Milton M. Zaret, a well-known expert on eye defects induced by microwave radiation, has pointed out that the increased incidence of coronary artery disease in cities may be related to the increased ambient level of electronic smog in the same cities. Studies in Russia have concluded that the central nervous system is susceptible to microwave radiation.

Despite the lack of complete data, basic safety limits have been established for microwave and radio wave exposure.

The American limit was established by the American National Standards Institute in November 1966 in Standard C-95.1, 1966, entitled *Safety Level of Electromagnetic Radiation with Respect to Personnel.* This standard sets a power density level of 10 milliwatts per square centimeter ($mW/cm^2$) for exposure times greater than six minutes and an energy density limit of one milliwatt hour per square centimeter ($mWh/cm^2$) for periods less than six minutes. The energy-density concept is a time-weighted exposure criterion by which the allowable exposure-time-in-hours-per-0.1 hour may be determined by dividing 1 $mWh/cm^2$ by the incident power density, expressed in $mW/cm^2$. For a power density of 60 $mW/cm^2$, as an example, the allowable exposure time is 1 $mWh/cm^2$ divided by 60 $mW/cm^2$, which is $\frac{1}{60}$ hr or 1 minute per 0.1 hour. (Excerpted from *A Review of International Microwave Exposure Guides*, by Jon R. Swanson, Vernon E. Rose, and Charles H. Powell, Bureau of Radiological Health Publication BRH/DEP 70-26.)

When moderate-to-severe heat stress is involved the limits should be appropriately reduced. And, under conditions of intense cold, higher limits may be appropriate in certain situations. At this time, certain modifications to these limits may be in order. Somewhere below 100 MHz the flux standard of 10 $mW/cm^2$ becomes unnecessarily conservative, in the opinion of Herman P. Schwan, who was instrumental in establishing some of the earlier radiation limits. Schwan has suggested that protection guide limits be set as follows: 3 $mA/cm^2$ for frequencies above 10 MHz, 1 $mA/cm^2$ for frequencies from 10 kilohertz (kHz) to 10 MHz, and 0.3 $mA/cm^2$ for frequencies below 10 kHz (data taken from *IEEE Transactions on Biomedical Engineering*, July 1972 © IEEE).

In addition to biological effects of nonionizing radiation, there is the serious problem of interference with heart pacemakers and the problem of accidental detonation of explosives.

## ELECTROEXPLOSIVE DEVICES

The following discussion is excerpted from Air Force T. O. 31Z-10-4.

Electroexplosive devices (EEDs) such as blasting caps, igniters, squibs, and primers are used in a variety of military and civilian applications. Military uses include conventional ground, ship, and aircraft ordnance as well as the more exotic missiles and rockets. Civilian uses include mining and wrecking plus newer applications in forming, shaping, perforating, and riveting metal. In any application of EEDs, the user must guard against the effects of lightning, static electricity, stray energy, and electromagnetic energy. The following paragraphs describe some of the basic hazards created by radio frequency (RF) energy.

RF energy can accidentally ignite EEDs even though the device is grounded or shorted. The probability of such a disastrous detonation depends on many factors, including the power output and frequency of radiation of nearby transmitters, directional characteristics of antennas, distance between the antenna and the firing circuit lead wires, shielding of the EED, and firing circuit configurations that are susceptible to resonance and induced currents. Whether or not a specific EED setup is hazardous may be difficult to determine; therefore, until experts have thoroughly investigated a particular situation, it must be assumed that a hazard exists whenever an EED is exposed to a high-intensity RF field.

A typical EED—the squib—is shown in Figure 5-5. It consists essentially of a flammable material in contact with an electrical transducer.

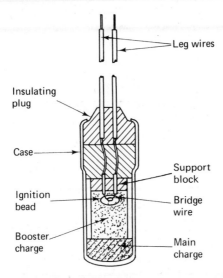

**Figure 5-5.** Typical electric squib. (Courtesy of USAF. From Air Force T.O. 31Z-10-4.)

Regardless of its shape or size, it usually consists of four major components: the explosive material, electric leads, a heat-sensitive filament or bridge across the leads, and a metal cylinder which encloses these items.

The basic firing circuit for a squib includes all circuits and components between the power source and the explosive element of the squib. The wiring in this circuit presents the greatest problem in protecting the squib from accidental firing. Typical squib firing circuits are shown in Figure 5-6. In part (a) of the figure, a single, unshielded conductor is used to connect the squib to the firing source. However, since this conductor also acts as a receiving antenna for stray RF radiation, it should not be used.

Twisted-pair and parallel-pair wires, shown in parts (b) and (c) of the figure, are less likely to respond to RF, partly because they are insulated with high RF loss insulation. Probably the best arrangement is that shown in part (d). By enclosing twisted or parallel-pair wires in a single or double shield of copper braid, we can force RF currents to flow on the surface of the braid and not affect the firing circuit. Note that the shielding must also

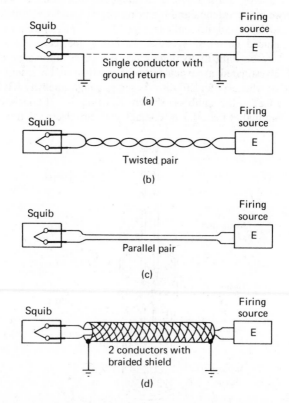

**Figure 5-6.** Typical squib firing circuits. (Courtesy of USAF. From Air Force T.O. 31Z-10-4.)

include the plugs, receptacles, switches, and the EED. The more complete the shielding, the greater the safety factor against accidental firing in the presence of high RF fields.

To establish a safe distance between an EED and the antenna of a nearby transmitter, consult *Safety Guide for the Prevention of Radio Frequency Radiation Hazards in the Use of Electric Blasting Caps*, ANSI C-95.4, available from the Institute of Makers of Explosives, 420 Lexington Avenue, New York, New York 10017, or local military directives if you are attached to a military base.

## COLOR TELEVISION

In the days of black-and-white TV, there was not enough high voltage in these sets to cause X-ray problems. With the advent of color TV however, high voltages—up to 25 kV—became common, and during 1966 and 1967 it was found that some of the large-screen color TVs across the nation were producing dangerous levels of X-rays. This radiation was found to originate in either the high voltage rectifier tube, the shunt regulator tube, the picture tube, or sometimes in all three (Figures 5-7 and 5-8). In most cases the primary cause of these X-rays was improper adjustment of the high voltage on the picture tube to levels higher than those recommended by the manufacturers. Secondary causes included excessive line voltage, design defects, and improper replacement of factory installed shields.

Fortunately, no injuries were caused by these excessive X-rays, and public concern soon forced both short- and long-term solutions to the problem.

Public Law 90-602, The Radiation Control for Health and Safety Act, was enacted in 1968 largely in recognition of the TV X-ray problem and other unnecessary exposure to radiation from all types of electronics products. One outcome of that law was the establishment of a performance standard for television receivers with the following key provisions for current models:

1. X-radiation exposure rates produced by a TV shall not exceed 0.5 milliroentgen (mR) per hour at a distance of 5 centimeters from any external point of the set.
2. All measurements shall be made: (a) with the set displaying a usable picture, and under the most adverse conditions: (b) with the power source operated at supply voltages up to 130 V rms, (c) with all user and service controls adjusted to combinations that result in the production of maximum X-radiation, and (d) with conditions identical to those which result from that component or circuit failure which maximizes X-radiation.

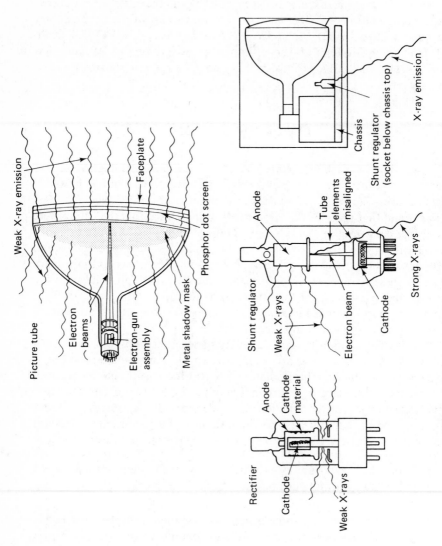

**Figure 5-7.** X-ray generators in a TV receiver. (Courtesy of © Popular Science Publishing Co., Inc., 1968.)

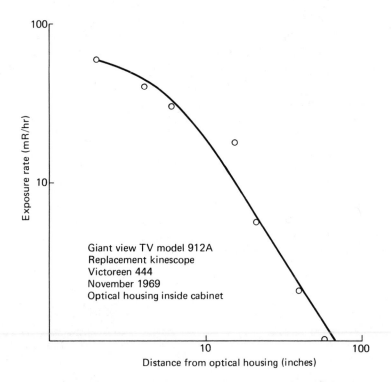

**Figure 5-8.** X-ray exposure from a projection TV. (From *Product Testing and Evaluation of Giant View TV Projector*, Bureau of Radiological Health 70-9, Department of Health, Education, and Welfare, May 1970.)

Properly operated and serviced TVs built since 1968 generally do not present a public health problem. As recently as January 1975, in one single recall, 400,000 color TVs were found to have excessive radiation. For TVs manufactured before June 1, 1971, however, it's possible that the first requirement above may not be met unless the supply voltage and the TV controls are properly adjusted. Therefore, the following safety precautions should be observed during service and operation:

1. Check picture tube high voltage with an accurate meter and compare with factory-recommended levels. Most conventional meters register 1 or 2 kV too low. (A calibrated electrostatic kilovoltmeter is ideal for accurate measurements.) During the check, turn the brightness and contrast controls to minimum (maximum illumination on picture tube screen).

2. Make sure that all factory-installed shields are in place and that the picture tube is a factory-specified type.

3. Check the line voltage to ensure that it does not exceed 130 V rms. If you are working on someone else's set, check the line voltage in the home at a time of day when the voltage is likely to be the highest.

4. Make sure that the high voltage rectifier tube and the high voltage regulator tube are of the latest lead-glass types.

5. Be aware that small screen color TV as well as large screen can generate X-rays. Black-and-white sets are not likely to produce X-rays.

6. Projection TVs are few in number but are much more dangerous than conventional TV.

7. Any viewing distance is satisfactory.

The Bureau of Radiological Health has designed two simple, inexpensive detectors (Figure 5-9) to indicate whether X-rays are being emitted from a TV at a rate that approaches the official limit of 0.5 mR/hr. These detectors car be built with parts commonly found in TV repair shops at a cost of about

**Figure 5-9.** Stoms-type large-area Geiger-Mueller survey instrument for measuring X-rays from color TV receivers. (Courtesy of Bureau of Radiological Health, Dept. of Health, Education, and Welfare.)

$25. (Plans may be obtained from the National Technical Information Service, Springfield, Va. for $3.00. Request BRH/DEP Publication 70-14, PB 192-377, *Simple X-ray Detection Instruments for Television Service Technicians.*)With these detectors, it is relatively easy to identify sets with X-radiation above maximum limits.

For those who prefer factory-built meters, Victoreen makes two radiation survey meters that can be used for detecting TV X-rays, the Model 499 "Vic-Chek" (Figure 5-10) and the Model 440 RF/C, which is recognized as the industry's standard. The Model 499 (Figure 5-11), which costs only one-tenth as much as the Model 440 RF/C, is used to detect X-radiation and to make a rough indication of whether the X-radiation is approaching the dangerous level. The Model 440 RF/C is used to measure accurately the exact extent of the problem.

The maximum permissible radiation exposure rate does not take into account the energy of the radiation being measured, although the radiation from a TV receiver may have a wide spectrum energy distribution. Consequently, a meter used to make accurate measurements of color TV X-rays must be energy independent. At these relatively low energy levels the most practical meter is one with an ion chamber and complex high impedance circuitry, such as the Model 440 RF/C. This model has been adopted by the JEDEC Committee of the Electronic Industries Association for low energy levels, but it is admittedly expensive.

Victoreen states that if a survey with a Model 499 produces no readings above midscale on the meter, you can be quite sure that the set is safe within

**Figure 5-10.** Victoreen Model 499 Vic-Chek x-ray monitor. (Courtesy of Victoreen Instrument Division.)

**Figure 5-11.** Victoreen Model 499 x-ray monitor in use. (Courtesy of Victoreen Instrument Division).

the present stipulated tolerance level. Should such a survey produce readings greater than half of full scale, Victoreen recommends that you inspect the set to ensure that:

1. The anode potential of the picture tube is at the level recommended by the manufacturer.
2. The high-voltage rectifier tubes and high-voltage regulator tubes are of the latest lead-glass types.

If an additional survey shows that X-radiation is being emitted at levels above midscale on the Model 499, the set should be surveyed with a Model 440 RF/C.

## KLYSTRONS AND HYDROGEN THYRATRONS

X-rays can be generated unintentionally under normal or abnormal operating conditions by klystrons, hydrogen thyratrons, magnetrons, and cathode ray tubes. X-rays created by such tubes may be relatively insignificant, hazardous, or extremely dangerous, depending on the voltage present in the tube. At 15,000 volts the accelerating potential produces relatively weak X-rays, which are not hazardous beyond a foot or so from the source; for such voltages, elaborate shielding is not necessary to protect nearby personnel. Some klystrons, however, operate with anode potentials up to 250 kilovolts (kV) and may accidentally produce high-energy X-rays that are 1,000 times as strong as X-rays deliberately produced in X-ray tubes. Adequate shielding is obviously essential at such levels.

Air Force T. O. 31Z-10-4 describes the klystron X-ray problem as follows:

> In effect, the high-velocity electrons bombard the collector assembly in a klystron in much the same manner as those in the X-ray tube bombard the target element. As a result, X-radiation occurs with greatest intensity in the region of the collector assembly. The shaded areas shown in Figure 5-12 illustrate the general distribution of X-radiation around a typical high-power klystron when it is in operation. Note that the greatest intensity is near the collector assembly, the output cavity, and the elbow bend of the outer waveguide. Radiation of lesser intensity occurs along the body of the tube approaching the output cavity and also from the electron gun itself.

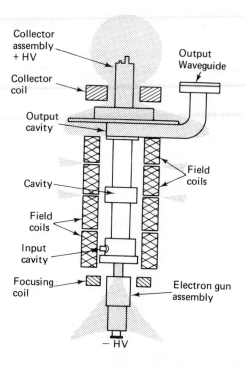

**Figure 5-12.** X-radiation distribution for typical high-power klystron. (Courtesy of USAF. From Air Force T.O. 31Z-10-4.)

X-radiation from the klystron occurs under two conditions: when the tube is in operation and delivering RF power to the load, and when high voltage is present but RF drive is not applied. In the latter condition, the electron beam travels through the tube to the collector, and X-radiation is produced predominantly in the collector region. However, when RF drive is applied to the tube and velocity modulation of the electron beam takes place, some dispersion of the electron beam causes emission from the body cavities of the tube. The increased acceleration of electrons near the output cavity region, caused by the higher effective voltage developed as a result of the applied dc potential and RF fields, gives rise to the generation of X-radiation of even greater intensity and penetrating power. Under normal operating conditions, the X-radiation measured in the immediate vicinity of the collector assembly may average 800 milliroentgens per hour for a typical klystron. Thus, as can be seen from the illustration in Figure 5-12, the klystron must be equipped with radiation shielding made of lead or other suitable material(s) installed over the collector, output cavity and main RF body, and electron gun assembly. In practice, the shielding is designed to attenuate the X-radiation to a level which is well within the maximum permissible level of 2.5 milliroentgens per hour established by USAF regulations. Without the radiation shields in place, the intensity of the radiation emitted is extremely dangerous to personnel.

Hydrogen thyratrons operating with high anode potentials may also emit X-rays. When thyratrons are used for high-voltage switching applications, as in radar pulse-modulator circuits, they may produce a considerable amount of X-radiation at the beginning of a pulse before the anode voltage drops, and a smaller amount of radiation between pulses. Radiation between pulses is caused, not by cathode electron emission, but by grid emission. The anode is surrounded by the grid structure in a thyratron tube, so that most of the radiation completely surrounds the tube in the form of a very narrow beam, extending outward from the grid anode region as shown in Figure 5-13.

*SAFETY PRECAUTIONS*

1. So long as the manufacturer's protective shielding remains intact on microwave tubes, there is no potential personnel hazard due to X-rays. Minor breaks and leakage points in the shielding, however, may permit X-rays to scatter into working areas. Thus, during routine maintenance or operation, the integrity of the tube shielding must be preserved to avoid exposure.

2. If the manufacturer's shielding must be removed during maintenance, it should be done only by technicians who are fully aware of the hazards involved. When servicing is complete, obviously the shielding must be replaced to prevent operating personnel from being unknowingly subjected to X-radiation.

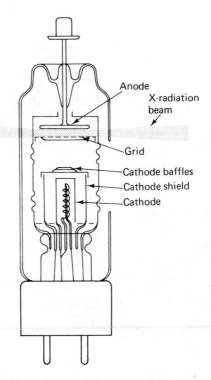

Anode
X-radiation beam

Grid

Cathode baffles
Cathode shield
Cathode

**Figure 5-13.** X-radiation distribution for typical hydrogen thyra-tron. (Courtesy of USAF. From Air Force T.O. 31Z-10-4.)

3. Adequate X-ray shielding must be provided when testing electronic devices that can produce X-rays, so as to protect all personnel in the test area.

4. Unless such procedures are called for in the instruction manuals, do not use jumper interlocks that allow equipment to be serviced with X-ray shielding removed.

5. Observe all instruction manual precautions, equipment warning signs, and local safety precautions, including the use of approved dosimeters.

6. In suspected areas of X-radiation, make dosimeter measurements with equipment cabinet doors closed and open.

The Victoreen Model 440 RF/C radiation exposure rate measuring system is specifically designed for X-ray leakage detection in the vicinity of electronics equipment that may have large electrostatic, magnetic, or electro-

**Figure 5-14.** Victoreen 440 RF/C ionizing radiation monitor. (Courtesy of Victoreen Co.)

magnetic fields associated with it (see Figure 5-14). The 440 RF/C is nonresponsive to such fields and responds only to ionizing radiation.

## COLD-CATHODE GAS DISCHARGE TUBES

Cold-cathode gas discharge tubes are used in demonstrations in many high schools and colleges (see Figure 5-15). These tubes may be divided into two basic types: those designed to produce X-rays and those that do not produce X-rays. In the latter group it has been found that heat-effect tubes, magnetic-effect tubes, and fluorescence-effect tubes (also known as shadow-effect or Maltese-cross tubes) can produce X-rays incidental to their intended use.

On the basis of laboratory and field studies, the Bureau of Radiological Health has come to the following conclusions about X-ray production from gas discharge tubes:

**Figure 5-15.** Cold-cathode gas discharge tubes. (Courtesy of the Bureau of Radiological Health, Dept. of Health, Education, and Welfare.)

1. X-ray output is sporadic; it may vary from one tube to another, and for the same tube from day to day.
2. Gas pressure within the tube is one of the controlling factors in X-ray production.
3. Tube construction plays an important part in the X-ray output obtainable from a tube.
4. The output of the tube is strongly dependent on the voltage and current capabilities of the power source and the polarity of the voltage.
5. It is impossible to predict whether a tube will produce X-rays, how long it will produce X-rays, or in what quantity.

Accordingly, the bureau has recommended the following precautions for the use of cold-cathode gas discharge tubes. Note that the recommendations apply to students less than eighteen years of age; students older than eighteen or teachers are not as restricted in their use of the tubes, but they would be well advised to follow the same precautions:

*Unshielded Cold-Cathode Gas Discharge X-ray Tubes.* The use of unshielded cold-cathode X-ray tubes in a classroom demonstration is not acceptable.

*Shielded Cold-Cathode Gas Discharge X-ray Tubes*

1. The maximum exposure rate at any point through the shielding should not exceed 10 mR per hour at a distance of 30 centimeters (cm) from the surface.

2. The primary beam from the beam port shall always be directed away from the class and the instructor.

3. No student shall use this equipment without appropriate on-the-spot supervision by the teacher.

4. No experiment should result in an exposure greater than 10 mR to any student.

*Gas Discharge Tubes Not Primarily Intended to Produce X-rays*

1. No student shall use this equipment without appropriate on-the-spot supervision by the teacher.

2. Tubes should always be operated at lowest current and voltage possible and time of operation kept to a minimum.

3. No experiment should result in an exposure greater than 10 mR to any student.

(The preceding information is from BRH/DEP 70-26, October 1970, *Radiation Emissions from Demonstration Type Cold-Cathode Gas Discharge Tubes*, by William S. Properzio.)

## MICROWAVE OVENS

Microwave, or "electronic," ovens (Figure 5-16) are a byproduct of World War II radar development and were placed on the market in 1947. However, it took twenty more years of cost and size reduction to make them

Figure 5-16. A hazard-free oven: Amana Radarange ®. (Courtesy of Amana Corp.)

a popular supplement to the standard cooking oven, rather than an expensive luxury.

The basis for this popularity is obvious—the amazing speed with which these ovens cook, taking typically one-third or less the time required by conventional ovens. Additional benefits include: a cooler kitchen, inasmuch as the oven does not get hot; less dishwashing, as many foods may be cooked in or on their serving dishes; and economy—electric power bills are much lower.

Such features are causing a revolution in cooking and have made microwave ovens useful in restaurants and institutions, as well as in the home. Aside from food preparation, the ovens are also being used in such diverse industry applications as sterilization, warming blood, and curing photo resist for printed circuits.

Most of the home models operate off standard 120-volt, 15-amp power lines and require no special installation except to ensure proper grounding.

Standard components of the microwave oven include a magnetron operating at 2450 megahertz (MHz), a resonant cavity which serves as the cooking compartment, and a waveguide to feed the microwave energy from the magnetron to the cavity. A fan-type stirrer (Figure 5-17) helps distribute the energy evenly throughout the cooking compartment. As cooking temperature is not a factor in electronic cooking, there is no temperature control on the oven. However, timing controls are provided for precise control of the cooking interval.

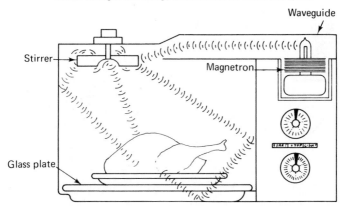

An S-band magnetron radiates microwave energy into oven cavity via a length of rectangular waveguide. The stirrer changes the energy distribution to avoid hot spots.

**Figure 5-17.** Action of microwave oven stirrer. (From Dan. R. McConnell, "Microwave Ovens—Revolution in Cooking," *Electronics World*, Aug. 1970, Copyright 1970 by Ziff-Davis Publishing Company.)

In conventional cooking, heat is applied to food from the outside. Through the process of convection and conduction, the food cooks from the "outside in." In microwave cooking, the microwaves penetrate and cook the entire depth of the food simultaneously. The air around the food is not heated. As the microwave energy is absorbed by the food, heat is quickly produced throughout the food, although there is less heat at the center than at the outside. This heat is produced by the action of the food molecules, which try to align themselves with the electric field generated by the microwaves. With this field alternating millions of times per second, the molecules are in a constant state of rotation. Because of molecular friction encountered during this rotation, heat is generated and distributed to the surrounding molecules.

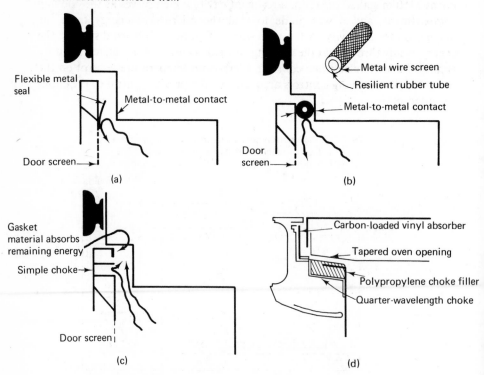

Various door seals designed to contain the microwave energy and prevent leakage. (a) Simple metal-to-metal contact using metal spring strip. (b) Compressible tube covered with metal screen or braid. (c) Simple quarter-wave choke seal puts short circuit on opening and prevents leakage. (d) A choke-absorber door seal acts to attenuate harmonics as well.

**Figure 5-18.** Types of microwave oven door seals. (From Dan R. McConnell, "Microwave Ovens—Revolution in Cooking," *Electronics World*, Sept. 1970, Copyright 1970 by Ziff-Davis Publishing Company.)

Paper plates, glass, and ceramics make good microwave cookware because they are generally transparent to microwaves and do not absorb energy.

Tests by the Bureau of Radiological Health have shown that some microwave ovens built before 1971 could be hazardous because of the microwave leakage they produce. Poor design, physical abuse, poorly adjusted interlocks, and damaged door seals (Figure 5-18) have been some of the contributing factors. Regardless of the cause, it should be noted that adverse effects from exposure to the radiation have not been confirmed.

The typical magnetron in a microwave oven operates at a moderately high voltage—from 2,000 to 7,000 volts. While this voltage is an obvious shock hazard, it is not high enough to produce any detectable X-radiation. X-rays simply are not a problem with microwave ovens. Any radiation hazard that may exist comes from the 600 or more watts of microwaves produced in the cooking process.

By law[2] the maximum allowable leakage radiation from a microwave oven is now 1 mW/cm², measured 5 cm (2 in.) from any of the oven's outer surfaces, at the time of manufacture. During the oven's useful life, this level may not exceed 5 mW/cm² no matter how much the doors and seals may deteriorate.

Since microwave power varies inversely as the square of the distance, microwaves lose their power fast (see Figure 5-19). Radiation of 1 mW/cm²

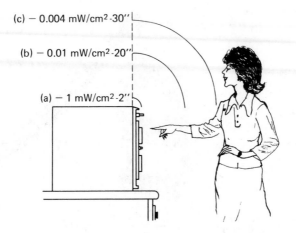

(c) — 0.004 mW/cm²-30″
(b) — 0.01 mW/cm²-20″
(a) — 1 mW/cm²-2″

**Figure 5-19.** Microwave-radiation power densities at various distances. (From Dan R. McConnell, "Microwave Ovens—Revolution in Cooking," *Electronics World*, Sept. 1970, Copyright 1970 by Ziff-Davis Publishing Company.)

[2]*Performance Standard for Microwave Ovens*, Federal Register, Vol. 35, No. 195, Title 42, Part 78, pp. 15619–15666 (Oct. 6, 1970).

at 2 in. from an oven, for example, becomes 4/1000 mW/cm² at 30 in. Thus, the *new* ovens pose no radiation hazard in normal use. (Hazards during repair are an entirely different matter.)

To maintain a low level of radiation, oven manufacturers have installed two safety interlocks in each oven to keep the oven from accidentally radiating when the oven door is open. Door seals have been designed, theoretically at least, to prevent microwave leakage even after years of use and abuse. Despite these safety features, it is possible, under rare conditions of oven damage and misuse, for some (but not all) microwave ovens to be hazardous, especially to the technician who may be repairing them. Thus, the oven technician should know the dangers involved and how to make leakage measurements.

*LEAKAGE MEASUREMENT*

The Narda Model 8100 Electromagnetic Leakage Monitor has long been the standard detection instrument for determining the amount of energy being leaked by microwave ovens. The newer Model 8200 (Figures 5-20 and 5-21) is simpler and less expensive than the 8100, yet provides all the essential features of the 8100. Like the 8100, it is a completely portable unit and has two interchangeable probes, which permit the selection of four ranges of

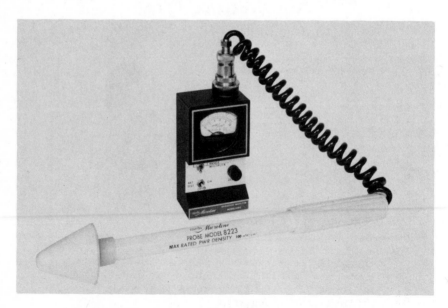

**Figure 5-20.** Narda Model 8200 radiation monitor. (Courtesy of Narda Microwave Corp.)

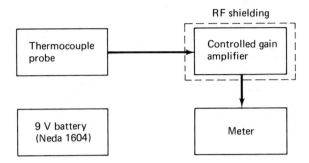

**Figure 5-21.** Narda Model 8200 electromagnetic leakage monitor overall simplified block diagram. (Courtesy of Narda Microwave Corp.)

TABLE 5-5  *Narda model 8200 specifications*[1]

| | |
|---|---|
| Calibration Frequency: | 2450 MHz |
| Power Reading Ranges: | Full scale (two ranges): |
|   Model 8223 | 10 mW/cm$^2$ and 100 mW/cm$^2$ |
|   Model 8221 | 2 mW/cm$^2$ and 20 mW/cm$^2$ |
| Accuracy of Thermocouple | ± 1 dB |
|   Probe Calibration: | |
| Thermocouple Probe Time | 0.5 second |
|   Constant: | |
| Response Time* | 1.2 |
|   (Nominal) | |
| Accuracy of Instrumentation: | ±3% of full scale |
| Thermocouple Probe | |
|   Overlock Rating: | |
|   Model 8223 | 300 mW/cm$^2$ |
|   Model 8221 | 40 mW/cm$^2$ |
| Meter Scales: | Linear Scale, marked 0–2 and 0–10 |
| Battery: Type Life | 9 volt NEDA 1604 |
|   (Approx.) | 40 hours |
| Size: | |
|   Thermocouple Probe | 11 inches long, 3/4 inch diameter |
|   Meter | 1–3/4″ × 4–5/8″ × 2–5/8″ |
|   Cable—Coiled | 6 inches long coil |
|     Uncoiled | 43 inches long coil |

*Response time including meter (the time it takes for the meter indicator to reach 90% of its final steady state reading when subjected to a stepped input signal).

[1]Operation and Maintenance Manual for Model 8200 Electromagnetic Leakage Monitor. (Reprinted by permission of the Narda Microwave Corporation.)

power density measurements. Specifications for the unit are given in Table 5-5.

In making leakage measurements, observe the following precautions:

1. When measuring an unknown field, start as far as possible from the source.

2. Do not position yourself so that your eyes are in a direct line with the source of the microwave energy.

3. If you do not know the approximate power density of the field to be measured, use the thermocouple probe with the highest power rating. If the power density is known, use the probe having the low power range equal to or above the known power density.

In all tests keep a standard load—275 milliliters (ml) of water—in the center of the oven. Hold the probe perpendicular to the oven surfaces (Figures 5-22 and 5-23) and pass the probe around the door seal (the most likely source of radiation). Leakage measurements should be made with the door closed and with it partially open (to check operation of interlocks). Specific measurement instructions are provided with the Model 8200.

**Figure 5-22.** Model 8100, *Narda Surveyor* in use to check radiation at microwave oven. (Courtesy of Narda Microwave Corp.)

*SAFETY PRECAUTIONS*

During installation, operation, or servicing of a microwave oven, follow the procedures recommended by the oven manufacturer. In addition, the following precautions may help you to avoid radiation exposure or electric shock in case the age of the oven is unknown. (Many of these procedures may not be necessary for ovens manufactured after October 1971.)

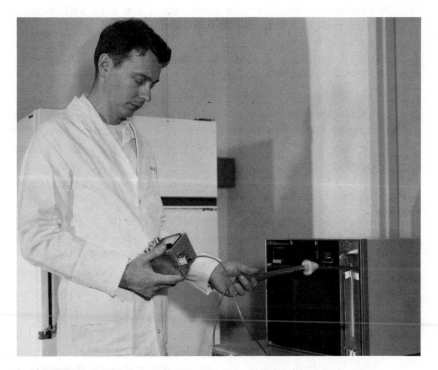

**Figure 5-23.** Testing microwave oven for leakage. (Courtesy of the Bureau of Radiological Health, Dept. of Health, Education, and Welfare.)

*Installation*
1. Examine the oven for shipping damage: broken, loose, cracked, or misaligned parts. Such defects must be repaired before the oven is used.
2. If the oven has been dropped, do not attempt to operate it until extensive visual checks are made.
3. Make sure the oven is properly grounded.

*Operation*
1. Never operate an empty oven.
2. Do not insert objects through the door grill (viewing window) or try to force objects between the oven and the door seal.
3. For ovens whose doors swing down, avoid setting heavy dishes on the open door.
4. Avoid leaning on the oven door.

5. Frequently clean the oven cavity, door, and seals with water (and mild detergent, if manufacturer approves). Dirt buildup on door seals can cause dangerous leakage. Use water sparingly so as to keep out of the waveguide. If stirrer is accessible and manufacturer recommends cleaning it, use a minimum of force.

6. Do not use scouring pads, steel wool, or other abrasives.

7. While cleaning, look for cracked door seals, grease buildup on seals, warped or misaligned door, and loose or broken hinges and latches.

8. For ovens built before October 6, 1971, turn the oven OFF before opening the door.

9. Stay at least an arm's length away from the front of the oven while it's ON.

10. Do not allow any object, even paper towels, to become trapped in door seals.

11. Never tamper with or inactivate the oven safety interlocks.

12. Follow specific safety instructions included in user and service manuals.

*Servicing*

1. Do not rely on the interlocks to cut off the oven. High voltage (shock hazard) may still be on if interlocks are defective or malfunctioning. Also, with high voltage on, the magnetron may be producing high-level radiation.

2. Do not modify the oven.

3. Avoid blows to the permanent magnets on those magnetrons so equipped.

## RADAR

Under certain conditions, as most technicians and engineers know, pulse and cw (continuous wave) radar can cause biological injury to anyone unlucky enough to be exposed to the radar's beam. Whether a particular radar is dangerous or not depends not only on the power and frequency (see Table 5-6), but also on the following factors:

1. Beam height—the main beam may be as much as 40 to 120 ft above the ground, simply because the radar is built above ground. Personnel in nearby low buildings, therefore, are protected. Buildings also provide protection even in direct beams (see Table 5-7).

2. Antenna azimuth and elevation angles; the customary positive tilt raises the beam above personnel on the ground.

TABLE 5-6  *Radar band designations*[1]

| Band Designation | Frequency (MHz) | Wavelength (cm) |
|---|---|---|
| P | 220–390 | 133.3–76.9 |
| L | 390–1550 | 76.9–19.3 |
| S | 1550–5200 | 19.3–5.77 |
| C | 3900–6200 | 7.69–4.84 |
| X | 5200–10,900 | 5.77–2.75 |
| K | 10,900–36,000 | 2.75–0.834 |
| Q | 36,000–46,000 | 0.834–0.652 |
| V | 46,000–56,000 | 0.652–0.536 |

[1]Source: Air Force T. O. 31Z-10-4.
(Reprinted by permission of the USAF.)

TABLE 5-7  *Microwave attenuation for various materials*[1]

| Wood Frame Building | |
|---|---|
| **Frequency (MHz)** | **dB Down (Power)** |
| 1300 | 2.0 |
| 2800 | 3.1 |
| 9200 | 1.3 |

| Interior Finishing Materials | |
|---|---|
| **Material** | **dB Down (Power)** |
| Plain tar paper (lightweight) 0.074″ thick | 2.5 |
| Pressed cardboard 0.351″ thick | 2.8 |
| Plaster board 0.362″ thick | 1.0 |

| Cinder-Block Wall | |
|---|---|
| **Frequency (MHz)** | **dB Down (Power)** |
| 1300 | 11.4 |
| 2800 | 14.5 |
| 9200 | 20.5 |

| Microwave Absorbent Material | | |
|---|---|---|
| **Frequency (MHz)** | **dB Down (Power)** | |
| | (wet) | (dry) |
| 1300 | 13.6 | 13.8 |
| 9200 | 25.0 | 25.0 |

| Copper Wire Screen | |
|---|---|
| **Frequency (MHz)** | **dB Down (Power)** |
| 1300 | 20 |
| 9200 | 24 |

[1]Source: Air Force T. O. 31Z-10-4.
(Reprinted by permission of the USAF.)

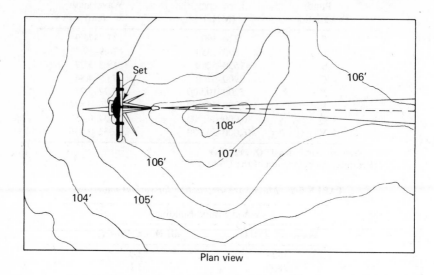

Plan view

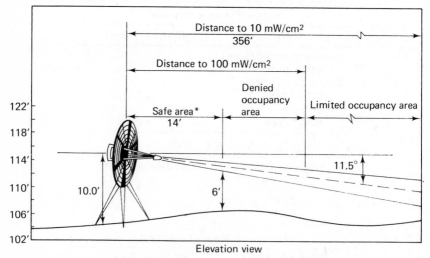

Elevation view

*Radiated beam is inaccessible to personnel at ground level to a distance of 14' from antenna.

**Figure 5-24.** Hazardous radar site conditions; example: Hawk high-power illuminator. (Courtesy of USAF. From *Control of Hazards to Health from Microwave Radiation*, Air Force AFM 161-7, Dec. 1965.)

3. Terrain and site configuration (see Figure 5-24).
4. Personnel duties—do normal job duties require personnel to enter areas that might be hazardous?

5. Whether the beam is stationary or is being rotated or scanned. Obviously, there is less radiation exposure if any antenna beam is scanning or rotating instead of radiating continuously in a stationary beam. In some cases, the on-axis power density of a radar may exceed the safe limit for continuous exposure, yet the radar will not be considered hazardous as long as it is rotating.

At this time, radiation exposure limits (Figure 5-25) for radar workers are somewhat more liberal than in other fields, and they allow for consideration of the amount of time involved in the exposure. Supposedly radar personnel are well versed in RF hazards and could take control of any hazardous situation that might arise. Limits, in military areas at least, are a tradeoff between personnel safety and operational restrictions. In other words, you have to take chances when you're in a combat zone.

The Army and the Air Force have designated the following types of hazard areas near radars:

1. Nonhazardous Area—power density less than 10 mW/cm². Continuous exposure is allowed in this area.

2. Limited Occupancy Area—power density between 10 mW/cm² and 100 mW/cm²; a potential hazard area. Duration of exposure in this area is limited by the equation:

$$T_p = \frac{6,000}{W^2}$$

where $T_p =$ permissible time of exposure in minutes during any one-hour period.

$W =$ power density (in mW/cm²) in area to be occupied.

Note: This formula should not be applied to intensities greater than 55 mW/cm².

3. Denied Occupancy Area—power density greater than 100 mW/cm², a definite hazard. No admittance.

Both AFM 161-7 and NAVSHIPS 0900-005-8000 list numerous unclassified military radars along with the distances at which they are considered hazardous to personnel. These distances vary from 0 to 1,275 ft, the worst-case situation of a fixed (rather than a moving) beam. For radars not listed, theoretical calculations and power density measurements (Figure 5-26) may be used to determine the safe distance from radar antennas. Specific instructions for making hazard surveys are given in appropriate military manuals, including T. O. 31Z-10-4 and NAVSHIPS 0900-005-8000.

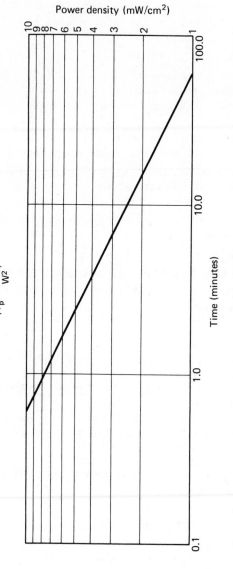

Power density (mW/cm²)

Permissible exposure time

$$T_p = \frac{600}{W^2}$$

Time (minutes)

Permissible exposure time may be obtained by use of figure.
Known power density = 25 mW/cm²
Enter figure on vertical axis at 25 mW/cm²
Follow horizontal entry line to intersection of curve.
Project vertical line from intersection to horizontal axis.
Read approximately 9.6 minutes.

Example:
Known: Power density = W = 25 mW/cm²

$$T_p = \frac{6000}{(25)^2} = \frac{6000}{625} = 9.6 \text{ min}$$

**Figure 5-25.** Permissible exposure time. (Courtesy of USAF. From *Control of Hazards to Health from Microwave Radiation*, Air Force AFM 161-7, Dec. 1965.

132

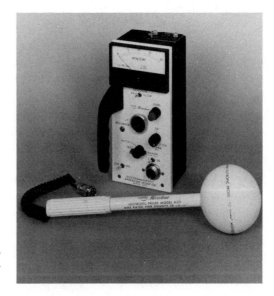

**Figure 5-26.** Narda Model 8315 electromagnetic radiation monitor. (Courtesy of Narda Microwave Corp.)

*SAFETY PRECAUTIONS*

1. Observe and heed RF HAZARDS warning signs. Do not enter DENIED OCCUPANCY AREAs unless the necessary transmitters have been turned off and it is known that they will not be turned on again without notice. Enter LIMITED OCCUPANCY AREAS only for the permissible time interval.

2. Do not look into or inspect any device—feedhorn, open waveguide, reflectors, radiators, etc.—while power is on.

3. Do not climb towers or other structures in the region of a high-intensity RF field until you have verified that the transmitter(s) have been turned off.

4. Use fencing or other structures to keep nonoperation and test personnel out of danger areas. If this is not possible, use bells, horns, flashing lights, public address system announcements, etc., to warn personnel of danger.

5. Erect RF screens or shields to protect personnel who must walk near hazardous radars.

6. During test operations, provide an enclosed dummy load (such as an absorbent material or a water load). Insofar as possible, direct non-rotating antennas away from inhabited areas while they are radiating.

7. If you work around radar, get a complete eye examination before going to work; then have your eyes checked every three years for radiation damage. The Navy's Bureau of Medicine and Surgery recom-

mends that you have your eyes checked immediately if you are exposed to more than 50 mW/cm² radiation.

## BROADCAST AND COMMUNICATIONS TRANSMITTERS

Public exposure to high-power RF or microwave radiation is not limited by any government or private standards. Yet people in or near a large city may be subjected to radiation from hundreds of radio and TV broadcast stations, two-way radios, and airport and military radars. In 1969, within a fifty mile radius of the Washington Monument it was found that there were 1,430 communications sources and 99 radar sites! And this total included only those unclassified transmitters with more than ten watts average output.

To determine if this radiation constituted a hazard, the Bureau of Radiological Health instituted a program of measurement of man-made radiation levels at ten scattered sites within twenty-five miles of Washington, D. C., during the summer of 1969. The following power density levels were observed:

| Frequency (MHz) | Power density exposure (mW/cm²) |
|---|---|
| Less than 400 | $3.9 \times 10^{-4}$ |
| 400 to 1,000 | $1.1 \times 10^{-5}$ |
| 1,000 to 3,000 | $7.7 \times 10^{-3}$ |
| 3,000 to 10,000 | $1.4 \times 10^{-4}$ |

Because these power levels were much less than any published U. S. recommendation for exposure limits, the Bureau concluded that no further broadband environmental surveys needed to be carried out unless prior

**TABLE 5-8** *Power levels for broadcast stations*[1]

**Maximum powers, tower heights, and ground-level field intensities for various broadcast services estimated at one-mile upper limits**

| Service | Maximum Allowable ERP, kW | Tower Height, meters | Field Intensity, mV/m |
|---|---|---|---|
| FM radio | 100 | 152.4 | 1023 |
| VHF television | | | |
| Channels 2–6 | 100 | 304.8 | 807 |
| Channels 7–13 | 316 | 304.8 | 191 |
| UHF television | | | |
| Channels 14–83 | 5000 | 304.8 | 380 |

[1]Richard A. Tell, "Broadcast Radiation: How Safe Is Safe?" *IEEE Spectrum*, August 1972. (© IEEE.)

calculations indicate their need. However, the Bureau pointed out that individual transmitters should be closely monitored to determine exposure levels in nearby populated areas.

Richard A. Tell of the U. S. Environmental Protection Agency notes that in 1971 there were 7,868 broadcasting stations on the air. In his paper, "Broadcast Radiation: How Safe Is Safe?" (*IEEE Spectrum*, August 1972),

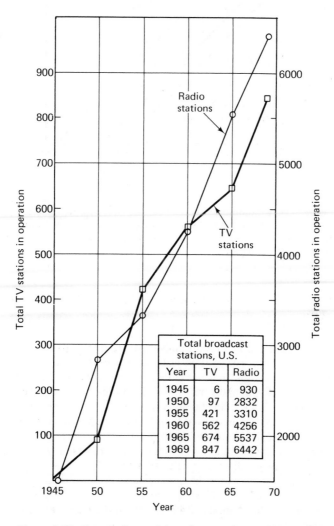

| Total broadcast stations, U.S. | | |
|---|---|---|
| Year | TV | Radio |
| 1945 | 6 | 930 |
| 1950 | 97 | 2832 |
| 1955 | 421 | 3310 |
| 1960 | 562 | 4256 |
| 1965 | 674 | 5537 |
| 1969 | 847 | 6442 |

**Figure 5-27.** Growth Rate of broadcast stations, 1945 to 1970. (Courtesy of the © *IEEE*. From Richard A. Tell, "Broadcast Radiation: How Safe Is Safe?," Aug. 1972, *IEEE Spectrum*.)

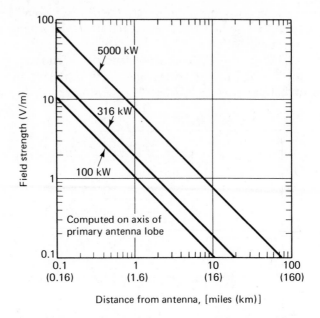

Figure 5-28. Relationship between field strength, effective radiated power, and distance from antenna. (Courtesy of the © *IEEE*. From Richard A. Tell, "Broadcast Radiation: How Safe Is Safe?," Aug. 1972, *IEEE Spectrum*.)

Tell makes the following observations (see Table 5-8 and Figures 5-27 through 5-29):

> Although fences usually prevent individuals from trespassing inadvertently on a station's property, often just the area near the tower base is fenced, and so it is possible to get as close to the tower base as 7.5 meters, where very high field strengths may exist. Exposures might approach 40 V/m at 160 meters from a directional AM station. This figure compares with 20 V/m, the maximum allowable full-time exposure limit for these frequencies in the Soviet Union and 10 V/m in Czechoslovakia.
>
> Potentially, worst-case situations could occur with persons working or living in tall buildings adjacent to television or FM installations where they would be placed in or near the primary lobe of the antenna beam. In the case of a 5-MW ERP, UHF station at 160 meters, the field strength could be as high as 76.1 V/m or 1.54 mW/cm² in the main beam for the video carrier.
>
> Access to station property usually imposes reasonable restrictions on how close an individual can get to an FM or television antenna system. In

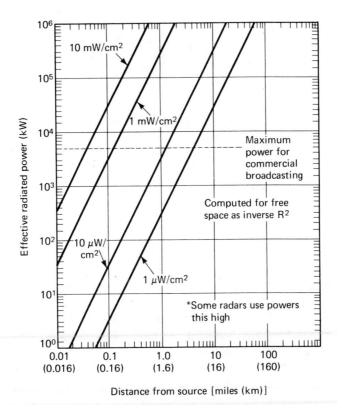

**Figure 5-29.** Effective radiated power versus distance required to produce various field power densities. (Courtesy of the © *IEEE*. From Richard A. Tell, "Broadcast Radiation: How Safe Is Safe?," Aug. 1972, *IEEE Spectrum*.)

addition, the tower height associated with these antennas and the high gain used offer an additional safety factor against high exposures.

Except for particularly unusual circumstances no normally encountered environmental levels realized from U.S. broadcast stations exceed any limits or standards anywhere in the world.

# 6

# LASERS

In recent years lasers have been put to work in diverse applications (Figures 6-1, 6-2, and 6-3), far removed from the strict controls of research laboratories. At the same time the cost of lasers has dropped to the point where even high school students can afford some models. With this widespread usage safety has too often been ignored, particularly with low-power lasers. Too many users have wrongly concluded that low-power (two milliwatt) lasers are not dangerous (see Table 6-1).

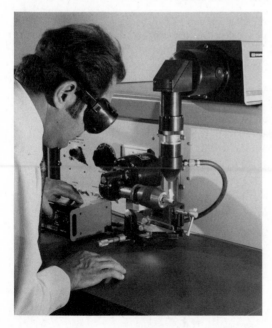

**Figure 6-1.** Laser welding machine. (Courtesy of GTE Sylvania Corp.)

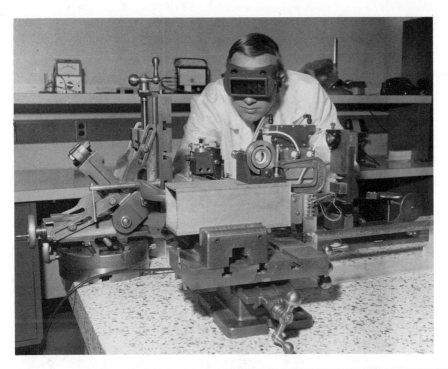

**Figure 6-2.** Note use of laser safety goggles (eyepiece). (Courtesy of Martin-Marietta Corporation.)

**Figure 6-3.** Dramatic flashes such as this require proper eye protection. (Courtesy of American Optical Corporation.)

**TABLE 6-1** Recommended limits for ocular exposure to laser radiation (0.4-1.4 μm) for a 7-mm pupil[1]

| Organization | Laser Wavelengths, nm | Exposure Duration $t$ | Intrabeam Viewing of a Collimated Beam | |
|---|---|---|---|---|
| | | | Corneal Radiant Exposure, $J \cdot cm^{-2}$ | Corneal Irradiance, $W \cdot cm^{-2}$ |
| U.S. Departments of the Army and Navy (February 1969) | 400–1400 | 5–50 ns | $10^{-7}$ | |
| | | Approx. 1 ms | $10^{-6}$ | |
| | | Continuous | | $10^{-6}$ |
| ACGIH (1971) | 694.3 | 1 ns to 1 μs | $10^{-7}$ | |
| | 694.3 | 1 μs to 0.1 s | $10^{-6}$ | |
| | 400–750 | >0.1 s | | $10^{-5}$ |
| U.S. Department of Labor 29 CFR 1518.54 (1971) | 632.8 | Indidental (1 s) | | $10^{-23}$ |
| | | Continuous | | $10^{-26}$ |
| U.S. Department of the Air Force (September 1971) | 400–700 | 10–100 ns | $1.3 \times 10^{-6}$ | |
| | | 200 μs to 2 ms | $10^{-5}$ | |
| | | 2–10 ms | | $5 \times 10^{-3}$ |
| | | 10–500 ms | | $2.5 \times 10^{-3}$ |
| | 1064 | 10–100 ns | $6 \times 10^{-6}$ | |
| | | 200 μs to 2 ms | $5 \times 10^{-5}$ | |
| | | 2–10 ms | | $2.5 \times 10^{-2}$ |
| | | 10–500 ms | | $1.3 \times 10^{-2}$ |

| ANSI Z-136 proposed (February 1972) | | | |
|---|---|---|---|
| 400–700 | 1 ns to 18 $\mu$s | $5 \times 10^{-7}$ | |
| | 18 $\mu$s to 10 s | $1.8 \times 10^{-3 \cdot t}$ | |
| | 10–$10^4$ s | $10^{-2}$ | |
| | >$10^4$ s | | $10^{-6}$ |
| 700–1060 | 1 ns to 18 $\mu$s | $5C_1 \times 10^{-7}$ | |
| | 18 $\mu$s to 10 s | $1.8C_1 \times 10^{-3 \cdot t}$ | |
| | 10–100 s | $C_1 \times 10^{-2}$ | |
| | $100 - [10^4/(\lambda - 699 \text{ nm})]$s | $C_1 \times 10^{-2}$ | |
| 700–800 | $>[10^4/(\lambda - 699 \text{ nm})]$s | | $C_1(\lambda - 699 \text{ nm}) \times 10^{-6}$ |
| | >100 s | | $C_1 \times 10^{-4}$ |
| 800–1060 | 1 ns to 100 $\mu$s | $5 \times 10^{-6}$ | |
| | 100 $\mu$s to 10 s | $9 \times 10^{-3 \cdot t}$ | |
| | 10–100 s | $5 \times 10^{-2}$ | |
| 1060–1400 | >100 s | | $5 \times 10^{-4}$ |

Note: $t$ is in seconds, $\lambda$ is wavelength in nanometers, and $C_1 = \exp\{[(\lambda - 700 \text{ nm})/224]\}$.

[1]Marce Eleccion, "Laser Hazards," *IEEE Spectrum*, August 1973.

(© IEEE)

Of the 30,000 or more lasers used in the nation's high schools and colleges, many have been used with inadequate safeguards, according to a seven-state survey by the Bureau of Radiological Health (FDA). Misuse appears to be common.

Although lasers are not *yet* the death ray of the future, assuredly they can blind you if they are used improperly. In addition, they can seriously damage exposed skin. The power supplies involved with lasers have dangerously high voltages—indeed, some researchers consider the laser power supply more hazardous than the laser itself.

If you work with lasers, you should have a complete ophthalmological examination before you are employed in a laser area. This exam should be repeated each year, and also at the time you stop working in a laser hazard area.

This chapter describes the properties of laser light, its biological effects, and safety precautions. Much of the following material has been abstracted from *Laser Fundamentals and Experiments*, by W. F. Van Pelt and others at the Southwestern Radiological Health Laboratory.

## PROPERTIES OF LASER LIGHT

As contrasted with the output of ordinary light sources, the characteristics of the laser's output are: small divergence, monochromaticity, coherence, and high intensity. These properties make the laser a valuable tool in many areas.

Because the light from a laser does not spread significantly, the energy in the beam is not greatly dissipated as the beam travels. The divergence of the beam is so small that it is measured in milliradians (1 milliradian equals about 3 minutes of arc). For a typical He—Ne laser, the rated divergence is 0.5–1.5 milliradians.

Laser light is almost monochromatic, that is, having one color or one wavelength of light. However, few lasers produce only one wavelength of light. A typical He—Ne laser, for example, emits light at 632.8 nanometers (nm), 1,150 nm, and 3,390 nm. The He—Ne laser is usually designed to emit only one of the three wavelengths of light and the variation in this wavelength is slight.

Two waves with the same frequency, phase, amplitude, and direction are said to be *spatially coherent*. Although no source of perfectly spatial coherent light is yet known, laser light is usually considered perfectly coherent. Only with sophisticated equipment is it possible to detect the variation from perfect spatial coherence.

Laser light can be very intense, in fact much brighter than the sun. Obviously, such intensity may be hazardous to your eyes.

The energy of a laser is commonly measured in joules (1 joule = 1 watt-second) or watts (1 watt = 1 joule/second). Thus, a 10-watt laser is one that can emit 10 joules in 1 second. If those same 10 joules are emitted as a single 1/100th second pulse, then the same laser may be termed a 1,000-watt laser.

Pulsed laser output is usually indicated in terms of joules per square centimeter ($J/cm^2$). The effect of the laser pulse depends upon the amount of time it takes to deliver the pulse. Consequently, pulsed laser output is sometimes referred to in terms of ($J/cm^2$)/sec or $W/cm^2$.

## BIOLOGICAL EFFECTS OF LASER LIGHT

The damage done to living tissue by laser light depends primarily upon the frequency of the light, the power density of the beam, the exposure time, and the type of tissue struck by the beam.

Damage may occur through thermal effect, acoustic transient, or other phenomena. The latter two effects only occur with high power density laser pulses.

When laser light strikes tissue, the absorbed energy produces heat, causing a rapid rise in temperature. The temperature rise can easily denature the protein material of tissue, much as an egg white is coagulated when cooked. The greatest thermal stress is around those portions of tissue that are the most efficient absorbers. If the absorption is rapid and localized, high temperatures may occur and there may be an explosive destruction of the absorber. At high exposure levels, steam may be produced, which can be quite dangerous if it occurs in an enclosed and completely filled volume such as the cranial cavity or the eye.

When laser light strikes tissue, part of its energy may be changed into a mechanical compression wave (acoustic energy), and a sonic transient wave can be built up. If it is near the surface, this sonic wave may send out a plume of debris from the impact; if beneath the surface, it can rip and tear tissue.

Free radical formation and other phenomena may exist during laser impact on biological systems, although this has not yet been conclusively demonstrated.

Usually the laser is a hazard to only those tissues through which the light beam can penetrate and which will absorb the wavelength involved. From a safety viewpoint we are primarily concerned with two organs: the eye and the skin.

To understand the hazard posed to the human eye by lasers, which may exceed the sun in irradiance, consider the cross-section of the eye as shown in Figure 6-4.

The tough white tissue forming the outer surface of the eye is called the *sclera*. The anterior portion of the sclera is specialized into the *cornea*, which

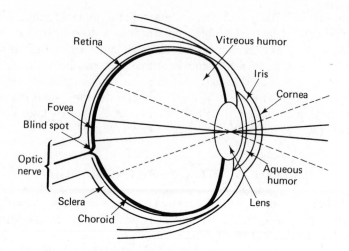

**Figure 6-4.** Schematic of human eye. (From W. F. Van Pelt et al., Southwestern Radiological Health Laboratory, Bureau of Radiological Health, Department of Health, Education and Welfare.)

is transparent to light and serves as the major focusing device of the eye. Within the eye, two fluid-filled cavities, both of which are under pressure, give structural rigidity to the eye. The anterior chamber contains a slightly viscous liquid, the *aqueous humor;* the rear chamber is filled with a very viscous, collagenous suspension, the *vitreous humor.*

These two chambers are separated by a lens attached by ciliary muscles to the sclera. These muscles alter the lens shape for fine focusing of the incoming light beam. Next to the lens is the pigmented *iris,* which expands or contracts to regulate the amount of light entering the eye.

Lining the rear fluid-filled chamber is the retina, which is composed of two tissues. The outer tissue contains the nerve cells for light perception. The underlying tissue stops light reflection, absorbs any scattered light, and provides support for the photoreceptor cells.

Before reaching the light sensor cells in the primate eye, light must first pass through several membranes, nerve fibers, ganglion cells, bipolar cells, and amacrone cells, and then must strike the photoreceptor cells from the rear.

The retina has two types of photoreceptor cells: rods and cones. Rods are quite sensitive to low light levels but cannot distinguish color. Cones are not as light sensitive but can distinguish color. In the retina, the two types are intermixed, with cones dominating near the center of the retina and rods near the periphery.

The *macula* (an area in the cones only) lies at the focal spot of the cornea-lens system. Within the macula is the *fovea,* a small region of densely packed cones which form the center for clear or critical vision. To one side of the

(a) Gamma and X-radiation
Most higher energy X-rays and gamma rays pass completely through the eye.

(b) Short ultraviolet
Absorption occurs principally at the cornea.

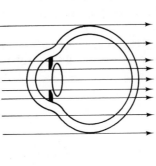

(c) Long ultraviolet and visible
Light is refracted at the cornea and lens and absorbed at the retina; long ultraviolet is absorbed on cornea and in lens.

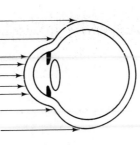

(d) Near infrared
Energy is absorbed in the ocular media and at the retina; near infrared rays are refracted.

(e) Far infrared
Absorption is localized at the cornea.

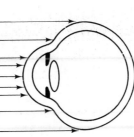

(f) Microwaves
Microwave radiation is transmitted through the eye although a large percentage may be absorbed.

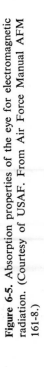

**Figure 6-5.** Absorption properties of the eye for electromagnetic radiation. (Courtesy of USAF. From Air Force Manual AFM 161-8.)

macula is a blind spot, at which point the nerve fibers from the photoreceptors exit the eye to form the optic nerve.

When light is focused by the cornea and lens onto the fovea of the retina, the energy density of the light is concentrated by a factor of $10^4$ to $10^6$ over that falling on the pupil. Because of this concentration, laser light may pose a serious hazard to the eye.

As shown in Figure 6-5, the human eye is relatively transparent to light at wavelengths from about 400 to 1400 nm, which includes not only the visible range of 400 to 700 nm, but also a portion of the infrared that is not perceived (Figure 6-6). The part of the eye affected by the laser depends on

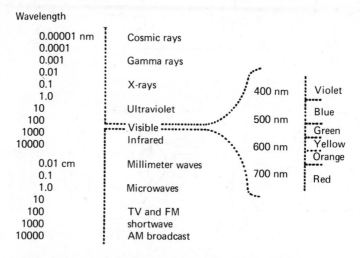

**Figure 6-6.** Electromagnetic spectrum. (From W. F. VanPelt et al., Southwestern Radiological Health Laboratory, Bureau of Radiological Health, Department of Health, Education and Welfare.)

the wavelength of the light. For example, the ruby laser emits at 694.3 nm. More than 90% of this light is transmitted through the ocular media to the retina. Of the light reaching the retina, about 60% is absorbed in the neuroectodermal coat, while almost all of the rest of the light is absorbed in the pigment epithelium. Since the pigment epithelium is only 10 micrometers thick, the greatest absorption per unit volume of energy occurs here, and this layer is the most susceptible to damage. Lesions may be produced here without the receptor cells being damaged.

Helium—neon, krypton, argon, and xenon lasers all operate in the visible range, and all affect the eye in the same manner as the ruby laser.

Neodymium laser light at 1060 nm, on the other hand, is absorbed more by the ocular media, with less of its energy reaching the retina than in the case of visible light. Thus there is a greater chance of damage by means of steam production than from other laser types. The aqueous and vitreous bodies are colloidal suspensions in water, and the absorption characteristics of the media are similar to those of water.

Carbon dioxide lasers produce light at 10,600 nm. At this wavelength, the eye is not very transparent and danger at low power densities is that lesions may be produced on the cornea.

Actually, power density *at the retina* cannot be measured, but must be calculated on the basis of transmission and focusing of the beam. The power density that can be measured is that *on the cornea*. On the basis of measurement at the cornea, lesions may theoretically be caused by as little as $10^{-6}$ J/cm² from a pulsed ruby laser.

Present threshold values for visible lesion production are approximately as follows:

| | |
|---|---|
| $Q$-switched ruby laser | @ 0.07 J/cm² on the retina |
| Pulsed ruby laser | @ 0.8   ″   ″   ″   ″ |
| Continuous white light | @ 6.0 W/cm² ″   ″   ″ |
| CO₂ laser | @ 0.2 W/cm² on the cornea |

Light levels below those producing visible lesions may also produce some permanent damage such as partial "bleaching" of the pigment for one particular light color.

Damage may result from laser impact on many different eye structures (see Figure 6-7). Oblique beam entrance may cause a lesion in the retina

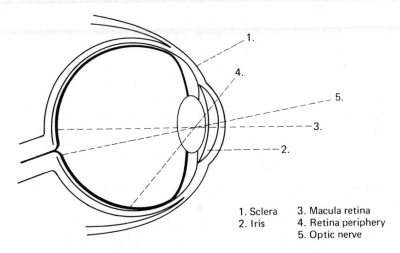

1. Sclera    3. Macula retina
2. Iris      4. Retina periphery
           5. Optic nerve

**Figure 6-7.** Possible sites of laser hit on eye. (From W. F. Van Pelt et al., Southwestern Radiological Health Laboratory, Bureau of Radiological Health, Department of Health, Education and Welfare.)

which goes unnoticed. A hit upon the optic nerve could result in complete blinding. The iris is dark colored and quite susceptible, while the whole sclera may fall victim to high energy beams.

The infrared emission of the $CO_2$ laser is destructive to the cornea. Corneal opacities are produced at a level of 0.2 $W/cm^2$ for 30 minutes continuous irradiation. As compared with other threshold levels, this is a high intensity, but then $CO_2$ lasers produce very intense beams and one is quite likely to encounter $CO_2$ lasers with continuous kilowatt outputs.

In calculating the maximum permissible exposure levels, the "worst case" hazard is assumed: the laser beam is assumed to be aimed directly at the fovea, the iris is dilated to produce a large pupil diameter, and the eye is focused at infinity. Under these circumstances, maximum irradiation of the retina wll occur.

*Skin Hazard.* The skin is not as sensitive as is the eye, and if damaged, it can often be repaired. However, when laser energy densities approach several $J/cm^2$, much skin damage is possible. Skin damage is probably not a hazard for He—Ne lasers. Laser damage to the skin ranges from a mild erythema to a surface charring, to a deep hole literally burned and blown into the skin.

The skin is a specialized, layered structure with numerous odd inclusions, such as blood vessels and hair follicles. As the skin is composed principally of water, laser beams interact as if skin were sea water containing a number of inclusions. As a result, the skin is relatively transparent to laser light.

In the skin, light absorption occurs mostly in the pigment granules and the blood vessels. The skin contains many pigments, the most common being melanin, which determine the color of one's skin. Visible laser light is selectively absorbed by the melanin granule, causing it to rise in temperature at a rapid rate, and causing cavitation around or bursting of the granule at high-energy density exposure.

Blood vessels are also quite susceptible to laser damage and are easily occluded or cauterized by a laser hit. Under certain circumstances, skin is so transparent that the visible light may pass through it to be absorbed by an internal organ.

## SAFETY PRECAUTIONS

The hazards posed by a laser can be determined accurately only by measuring several beam parameters. Unfortunately, such measurements require the use of calibrated and costly electronic gear. However, an approximate idea of these hazards may be gained by simply reading the manufacturer's specifications.

*CAUTION:* An individual laser may exceed the specifications listed by the laser manufacturer. A nominal 2-mW laser, for example, will usually have an actual output greater than 2 mW. A laser with a nominal divergence of 1.0 milliradians may have a divergence of 0.9 milliradians. Use of the nominal specifications to calculate the power density may lead to a serious underestimation of the hazards involved. However, figures supplied by manufacturers do provide a general idea of the power densities involved.

Safe or tolerable exposure levels are given in units of irradiance, mW/cm². Therefore, the output of your laser must be expressed in similar units. The usual information given by a laser manufacturer, and typical values for a classroom type He—Ne laser are given below:

| | |
|---|---|
| Power output | 1.0 milliwatt |
| Beam diameter at aperture | 1.5 millimeters |
| Beam divergence | 1.0 milliradian |

The irradiance at the aperture is given by the following formula:

$$P_a = \frac{P}{\text{area}} = \frac{P}{\pi(D_a/2)^2}$$

where $P_a$ = irradiance at aperture
$P$ = power output
$D_a$ = beam diameter at aperture in cm.

For the laser listed above

$$P_a = \frac{1.0 \text{ mW}}{3.14\,(0.15\,\text{cm}/2)^2} = 57.1 \text{ mW/cm}^2$$
$$= 5.7 \times 10^{-2} \text{ W/cm}^2$$

This laser, therefore, has an output irradiance of approximately $6 \times 10^{-2}$ W/cm² at the aperture. As compared with the "safe" value of $5 \times 10^{-5}$ W/cm² recommended by the American Conference of Government Industrial Hygienists (1968) for daylight illumination, it is approximately 1,000 times greater than the "safe" value.

As the laser beam travels beyond the aperture, it diverges slightly. At a distance of $r$ meters, the beam will have diverged to a diameter $D_r$, as shown in Figure 6-8.

The formula for determining the irradiance at a distance $r$ centimeters from the aperture is

$$P_r = \frac{P}{\pi(D_r/2)^2} = \frac{4P}{\pi(D_r)^2}$$

where $P_r$ = irradiance at distance $r$
$D_r$ = diameter of beam at distance $r$, in centimeters.

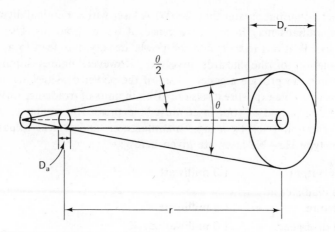

**Figure 6-8.** Beam divergence. (From W. F. Van Pelt et al., Southwestern Radiological Health Laboratory, Bureau of Radiological Health, Department of Health, Education and Welfare.)

The divergence of the beam in radians may be used to determine the beam diameter at distance *r* as follows:

$$D_r = r\phi + D_a$$

where   $r$ = distance in centimeters
   $\phi$ = divergence in radians
   $D_a$ = aperture diameter in centimeters.

Thus, the irradiance may be determined at any point. For example, the the irradiance at a distance of about 40 feet (13 meters)—a reasonable distance from the front to the rear of a classroom—would be:

$$D_r = (1,300 \text{ cm} \times 1.0 \times 10^{-3} \text{ radians}) + (0.15 \text{ cm}) = 1.45 \text{ cm}$$

$$P_r = \frac{4 (1.0 \text{ mW})}{3.14 (1.45 \text{ cm})^2} = 6.06 \times 10^{-4} \text{ W/cm}^2$$

Therefore, even a small classroom-type laser may emit enough power to be unsafe for direct viewing of either the primary beam or of specular reflection, even though the viewer may be seated in the rear of a classroom.

*General Safety Rules.* The primary hazard in laser work is overexposure of the eyes and the skin. Additional hazards include: electrical shock, toxic chemicals, exploding components (such as flash tubes or optics), cryogens for cooling, and noise or bright flash levels, which may startle personnel and thereby cause an accident. The last four hazards are not usually associated with He—Ne lasers.

The general rules given below should be followed whenever the laser is used in either a classroom or a laboratory:

### Work Area Controls

1. Keep the laser away from areas where the uninformed or curious would be attracted by its operation; that is, keep operations within closed areas posted with warning signs.
2. Keep illumination in the area as bright as possible in order to constrict the pupils of the observers. Room surfaces and equipment should be nonglossy to avoid reflections.
3. Set up the laser so that the beam path is not at normal eye level, that is, so it is below three feet or above $6\frac{1}{2}$ feet.
4. Use shields to prevent both strong reflections and the direct beam from going beyond the area needed for the demonstration or experiment.
5. Use a beam target that is a diffuse, absorbing material to prevent reflection. It should be fire resistant.
6. Remove all watches and rings before changing or altering the experimental setup. Shiny jewelry could cause hazardous reflection.
7. Cover all exposed wiring and glass on the laser to prevent shock and contain any explosions of the laser materials. Ground all nonenergized parts of the equipment. Use interlocks to prevent electric shock and laser light exposure.
8. Place signs in conspicuous locations both inside and outside the work area and on doors giving access to the area to indicate that the laser is in operation and that it may be hazardous. A flashing red light might be worthwhile.
9. Whenever possible, keep the door(s) locked to keep out unwanted onlookers during laser use. If they can't be locked, consider use of safety interlocks on doors.
10. Never leave a laser unattended.
11. Practice good housekeeping to ensure that no device, tool, or other reflective material is left in the path of the beam.
12. Prepare a detailed operating procedure beforehand for use during laser operation.
13. Install a warning device to indicate when a laser is operating outside the visible range (such as a $CO_2$ laser).
14. Install a key switch for the high voltage supply.

*Personnel Control*

1. Avoid looking into the primary beam *at all times.*
2. Do not aim the laser with the eye: direct reflection could cause eye damage.
3. Do not look at reflections of the beam: these too could cause retinal burns.
4. Avoid looking at the pump source at all times.
5. Clear all personnel from the anticipated path of the beam.
6. Do not depend on sunglasses to protect the eyes. If laser safety goggles are used, be certain they are designed to be used with the laser being used. (Table 6-2). Inspect safety goggles before each use. Determine failure point of goggles either from the manufacturer or test them at their highest intended power before using. (It's best to completely enclose the laser, the beam, and the target and not have to use safety goggles.)
7. Report any afterimage to a physician, preferably an ophthalmologist who has had experience with retinal burns, as damage may have occurred.
8. Be very cautious around lasers that operate at invisible light frequencies.
9. Before operation, warn all personnel and visitors of the potential hazard. Remind them that they have only one set of eyes.

TABLE 6-2 *Summary of AO Laser Eye Protection*

The table below shows the best choice of AO Laser Eye Protection for many common lasers. Each case, should, of course, be evaluated carefully to be sure of the best choice.

| Laser | Wavelength (nm*) | Goggle | Optical Density | Visual Transmittance |
|---|---|---|---|---|
| Ruby | 694 | 585 | 8 | 35% |
| Nd-doped glass, crystals, and liquids | 1060 | 584 | 11 | 46% |
| Argon | 455–515 | 598 | Min. 9 at 515 nm | 24% |
| | | 599 | Min. 8 at 515 nm | 25% |
| CO₂ | 10,600 | 680 | 50 | 100% |
| He-Ne | 633 | 581⎫ 587⎭ | 4 | 10% |
| GaAs | 700–1000 | 584 | Min. 4 at 700 nm Max. 14 at 900 nm | 46% |

*nm = nanometers = millimicrons (mμ)
(Courtesy of American Optical Company.)

10. Do not perform laser work alone—use the "buddy" system.
11. Where appropriate, provide exhaust hood over target.
12. If possible, turn face or keep eyes closed during laser operation.
13. When outdoors, treat the laser as an ordnance piece with similar security and safety measures.

Air Force manual AFM 161-8 gives these additional precautions:

*Viewing Beam Through Telescope.* The hazard of viewing a laser beam may be increased by viewing through a telescope. If the diameter of the beam is less than the pupil diameter, there is no change in the hazard. If the diameter of the beam at the telescope is less than the diameter of the objective lens, the entire beam energy or power must be considered to enter the eye. If the diameter of the beam at the telescope is larger than the objective lens, the energy or power entering the eye is increased approximately by the square of the magnification of the telescope. Telescopes, binoculars, theodolites, transits, or any optical viewing devices that magnify the image are considered to be telescopes in this discussion and may increase the hazard of viewing a laser beam. Viewing a noncollimated source through a telescope does not normally increase the hazard.

*Toxic Chemicals.* Toxic chemicals are sometimes found with liquid and chemical lasers, saturable dye $Q$-switches, Raman and Brillouin scattering cells, etc. Toxic gases are often produced as a result of high energy laser beams ionizing air or disintegrating the target.

*Ultraviolet and X-radiation.*
1. Ultraviolet and X-radiation are usually present with electric conduction through gases. Generally, X-rays are no problem if the peak voltages are less than 15,000 volts.
2. Ultraviolet radiation can produce toxic levels of ozone in the air and can produce keratoconjunctivitis and corneal injury if the eye is exposed. Ultraviolet radiation is relatively easy to shield. Most glass, except quartz, is relatively opaque to ultraviolet radiation, as are most plastics and construction materials that are opaque in the visible region of the spectrum. For this reason, gas discharge tubes should not be constructed from quartz glass if other materials will do the job.

*Cryogenic Materials.*
1. Cryogenic coolants, such as liquid helium and liquid nitrogen, are often used with laser systems to increase coherency or increase output power and are sometimes employed with detection devices. Liquid water cooling systems are preferable from a safety standpoint, but often cannot provide the low temperatures required by some systems.

2. There are many hazards associated with handling cryogenic liquids. Contact with the skin will produce a "burn," and great care must be exercised to avoid contact. It is recommended that asbestos gloves and a mask be worn when filling and pouring from Dewar and thermos bottles. The glass vacuum bottles, Dewar and thermos bottles should be protected by an outer container that will not permit flying glass to escape should the vacuum bottle implode. Containers should always permit blow-off of the gas as the cryogenic liquid evaporates.

3. Great care must be taken when cryogenic liquids are vented because air will be displaced by the vented gases, leaving an oxygen-deficient atmosphere (therefore, use adequate ventilation). Liquid helium and liquid nitrogen both will condense oxygen out of the air. Liquid oxygen and almost all hydrocarbons will combine to form a shock-sensitive explosive mixture. Hydrocarbons must not be permitted to cool to the point where liquid oxygen will condense on them, or placed where liquid oxygen may drip on them.

*Projectiles from Catastrophic Failure.* Due to the extremely rapid heating possible with high-powered laser beams, glass, crystalline, and other vitrified objects may shatter under thermal stress, producing dangerous projectiles. Laser rods and lenses have been known to disintegrate violently. Targets may release a shower of projectiles when hit with a high-powered laser beam. The target and elements of the optical train capable of shattering should be enclosed when used with high powered laser systems.

*Plume from Target.* Material ejected in the plume from the target may have various compositions, depending on the nature of the target. In addition to toxic gases, dusts, and fumes, it has been found that upon irradiation of cancerous tumors with laser pulses, numerous viable cells are ejected in the plume, many with genetic damage. Pathogens also may be ejected in the plume. Therefore, great care must be exercised to contain the plume and prevent contamination of the work space when potentially dangerous material may be ejected by a target.

*Environmental Control.* Lasers require air quality control (temperature, humidity, dust, etc.), as do most electrical and optical equipment. Environmental control is more critical with lasers because dust and condensate will cause serious degeneration of optical lenses and mirrors and can cause spurious reflections. Foreign material on lenses and mirrors leads to pitting and the hazard of ejected glass particles. Part of environmental control is good housekeeping. Good housekeeping eliminates unnecessary reflecting materials from the beam area and reduces the chances of accidental exposure to laser radiation from specular reflections. Obstacles on the floor pose a serious threat to personnel whose field of view is limited by laser safety eye wear. Good housekeeping and environmental control make the work area safer.

*GTE Sylvania recommends*

1. Energy sources for laser operations are essentially high voltage equipment and as such must be treated with caution. Electric shock and burns resulting from input power or capacitor discharge may cause serious injury or death.

2. Proper maintenance of cables, connectors, cabinets, and switches is essential. Discharge capacitors before cleaning or making repairs. Since capacitors can retain a charge even when the power is disconnected, provide bleeder resistors. Flash units once after the switch is turned off to remove most of the residual discharge from the capacitors. Where practical, install automatic discharge bars.

3. Low voltage power supplied for auxiliary instruments, measuring devices, etc., is a potential source of shock. Compliance with provisions of the State Electrical Safety Code will minimize these hazards.

4. Covers over high voltage circuits must be interlocked to prevent access to energized components. A tagout or lockout system must be used to provide assurance that all connections are made with power supplies disconnected.

5. Construction, operation, and repairs of power supplies and other laser components must be attempted only under supervision of a person familiar with the equipment and its potential hazards.

## PROPOSED LASER PERFORMANCE STANDARD

The Food and Drug Administration's Bureau of Radiological Health has proposed a safety standard to prevent unnecessary exposure to radiation from laser products. As of this time, it is to go into effect August 2, 1976.

Under this standard, laser products would be divided into four classes according to degrees of radiation risks. Class I products would permit exposure only to laser radiation at levels at which biological damage has not been established. Products that emit light that could cause eye damage after long-term exposure would be in Class II. Class III would cover laser products emitting accessible radiation levels at which damage to human tissue is possible from one short, direct exposure. Class IV would cover laser products which could cause biological damage after indirect as well as direct exposure.

Warning labels will be required for all but Class I laser products so that users will be aware of the hazard potentials of the devices prior to purchase. Labels for Class II devices (Figure 6-9) will have to caution: "LASER

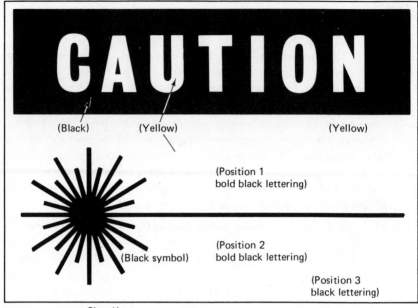

Class II
Position 1 — laser radiation — do not stare into beam
Position 2 — radiation output information
Position 3 — class II laser product

Class IIIa
Position 1 — laser radiation — do not stare into beam
    or view directly with optical instruments
Position 2 — radiation output information
Position 3 — class IIIa laser product

**Figure 6-9.** Proposed warning sign for Class II and IIIa lasers. (From the Federal Register, Vol. 40, No. 148, July 31, 1975.)

RADIATION—DO NOT STARE INTO THE BEAM." Class III product labels will have to warn against either direct exposure to the beam or staring into or viewing the beam with telescopes, binoculars, or other light amplification equipment. Class IV product labels (Figure 6-10) will have to warn: "LASER RADIATION—AVOID EYE OR SKIN EXPOSURE TO DIRECT OR SCATTERED RADIATION."

Protective housing and safety interlocks will be required to prevent user exposure to laser emissions in excess of the lowest levels at which products could perform their intended functions. Class III and IV lasers can be turned on only with a key.

Only Class I or II lasers may be used as demonstration lasers. Only Class I, II, or III lasers may be used for surveying, leveling, and alignment.

Accessible emission limits for laser radiation in each class are specified

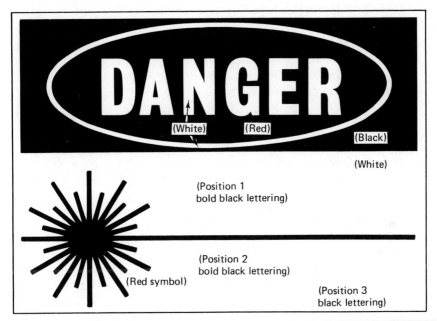

**Class IIIb**
  Position 1 — laser radiation — avoid direct
              exposure to beam
  Position 2 — radiation output information
  Position 3 — class IIIb laser product

**Class IV**
  Position 1 — laser radiation — avoid eye or skin
              exposure to direct or scattered radiation
  Position 2 — radiation output information
  Position 3 — class IV laser product

**Figure 6-10.** Proposed warning sign for Class IIIb and IV lasers. (From the Federal Register, Vol. 40, No. 148, July 31, 1975.)

in Tables 6-3, 6-4, and 6-5 in terms of the factors, $k_1$ and $k_2$, for different ranges of wavelength and emission duration. These factors are given in Table 6-6 with selected numerical values in Table 6-7, for various subranges of wavelength and sampling interval. The accessible emission limits in Tables 6-3, 6-4, and 6-5 are also expressed in terms of the sampling interval ($t$) for some ranges of emission duration. The correction factors in Table 6-6 are expressed in terms of the specific wavelength ($\lambda$) and sampling interval ($t$) for some subranges of wavelength and sampling interval. Accessible emission limits for collateral radiation[1] are specified in Table 6-8.

[1]Collateral radiation is any electronic product radiation, except laser radiation, emitted by a laser product as a result of or necessary for the operation of a laser incorporated into that product.

In regard to Tables 6-3, 6-4, and 6-5:

1. The quantities represented in the tables are radiant energy expressed in joules (J); radiant exposure expressed in joules per square centimeter ($Jcm^{-2}$); and integrated radiance expressed in joules per square centimeter per steradian ($Jcm^{-2}sr^{-1}$).
2. The factors $k_1$ and $k_2$ are wavelength dependent correction factors determined from Table 6-6.
3. The variable $t$ in the expressions of emission limits is the magnitude of the sampling interval in units of seconds.
4. An accessible emission limit containing the units of joules, when divided by the sampling interval($t$), is equivalent to an accessible emission limit containing the units of watts.

*Beam of a Single Wavelength.* Laser or collateral radiation of a single wavelength exceeds the accessible emission limits of a class if its accessible emission level is greater than the accessible emission limit of that class within any of the ranges of emission duration specified in Tables 6-3, 6-4, and 6-5.

*Beam of Multiple Wavelengths in Same Range.* Laser or collateral radiation having two or more wavelengths within any one of the wavelength ranges specified in Tables 6-3, 6-4, and 6-5 exceeds the accessible emission limits of a class if the sum of the ratios of the accessible emission level to the corresponding accessible emission limit at each such wavelength is greater than unity for that combination of emission duration and wavelength distribution which results in the maximum sum.

*Beam with Multiple Wavelengths in Different Ranges.* Laser or collateral radiation having wavelengths within two or more of the wavelength ranges specified in Tables 6-3, 6-4, and 6-5 exceeds the accessible emission limits of a class if it exceeds the applicable limits within any one of those wavelength ranges. This determination is made for each wavelength range.

*Class I Dual Limits.* Laser or collateral radiation in the wavelength range of greater than 400 nm but less than or equal to 1,400 nm exceeds the accessible emission limits of Class I if it exceeds both:

The Class I accessible emission limits for radiant energy within any range of emission duration specified in Table 6–3 and

The Class I accessible emission limits for integrated radiance within any range of emission duration specified in Table 6-3.

**TABLE 6-3**  *Class I accessible emission limits for laser radiation*

| Wavelength (nanometers) | Emission duration (seconds) | Class I—Accessible emission limits |
|---|---|---|
| > 250 but ≤ 400 | ≤ $3.0 \times 10^4$ | $2.4 \times 10^{-5} k_1 k_2$ J* |
| | > $3.0 \times 10^4$ | $8.0 \times 10^{-10} k_1 k_2 t$ J |
| > 400 but ≤ 1400 | > $1.0 \times 10^{-9}$ to $2.0 \times 10^{-5}$ | $2.0 \times 10^{-7} k_1 k_2$ J |
| | > $2.0 \times 10^{-5}$ to $1.0 \times 10^{1}$ | $7.0 \times 10^{-4} k_1 k_2 t^{3/4}$ J |
| | > $1.0 \times 10^{1}$ to $1.0 \times 10^{4}$ | $3.9 \times 10^{-3} k_1 k_2$ J |
| | > $1.0 \times 10^{4}$ | $3.9 \times 10^{-7} k_1 k_2 t$ J |
| | **OR**** | |
| | > $1.0 \times 10^{-9}$ to $1.0 \times 10^{1}$ | $10 k_1 k_2 t^{1/3}$ J cm$^{-2}$ sr$^{-1}$ |
| | > $1.0 \times 10^{1}$ to $1.0 \times 10^{4}$ | $20 k_1 k_2$ J cm$^{-2}$ *sr*$^{-1}$ |
| | > $1.0 \times 10^{4}$ | $2.0 \times 10^{-3} k_1 k_2 t$ J cm$^{-2}$ sr$^{-1}$ |
| > 1400 but ≤ 13000 | > $1.0 \times 10^{-9}$ to $1.0 \times 10^{-7}$ | $7.9 \times 10^{-5} k_1 k_2$ J |
| | > $1.0 \times 10^{-7}$ to $1.0 \times 10^{1}$ | $4.4 \times 10^{-3} k_1 k_2 t^{1/4}$ J |
| | > $1.0 \times 10^{1}$ | $7.9 \times 10^{-4} k_1 k_2 t$ J |

*Class I accessible emission limits for the wavelength range of greater than 250 nm, but less than or equal to 400 nm shall not exceed the Class I accessible emission limits for the wavelength range of greater than 1400 nm but less than or equal to 13000 nm with a $k_1$ and $k_2$ of 1.0 for comparable sampling intervals.

**Instructions for the Class I dual limits are set forth in paragraph (d)(4) of this section.
(From *Federal Register*, Vol. 40, No. 148, Thursday, July 31, 1975)

**TABLE 6-4**  *Class II accessible emission limits for laser radiation*

| Wavelength (nanometers) | Emission duration (seconds) | Class II—Accessible emission limits |
|---|---|---|
| > 400 but ≤ 700 | > $2.5 \times 10^{-1}$ | $1.0 \times 10^{-3} k_1 k_2 t$ J |

(From *Federal Register*, Vol. 40, No. 148, Thursday, July 31, 1975)

**TABLE 6-5**  *Class III accessible emission limits for laser radiation*

| Wavelength (nanometers) | Emission duration (seconds) | Class III—Accessible emission limits |
|---|---|---|
| > 250 but ≤ 400 | ≤ $2.5 \times 10^{-1}$ | $3.8 \times 10^{-4} k_1 k_2$ J |
| | > $2.5 \times 10^{-1}$ | $1.5 \times 10^{-3} k_1 k_2 t$ J |
| > 400 but ≤ 1400 | > $1.0 \times 10^{-9}$ to $2.5 \times 10^{-1}$ | $10 k_1 k_2 t^{1/3}$ J cm$^{-2}$ to a maximum value of 10 J cm$^{-2}$ |
| | > $2.5 \times 10^{-1}$ | $5.0 \times 10^{-1} t$ J |
| > 1400 but ≤ 13000 | > $1.0 \times 10^{-9}$ to $1.0 \times 10^{1}$ | 10 J cm$^{-2}$ |
| | > $1.0 \times 10^{1}$ | $5.0 \times 10^{-1} t$ J |

(From *Federal Register*, Vol. 40, No. 148, Thursday, July 31, 1975)

**TABLE 6-6** *Values of wavelength dependent correction factors $k_1$ and $k_2$*

| Wavelength (nanometers) | $k_1$ | $k_2$ |
|---|---|---|
| 250 to 302.4 | 1.0 | 1.0 |
| > 302.4 to 315 | $10\left[\dfrac{\lambda - 302.4}{5}\right]$ | 1.0 |
| > 315 to 400 | 330.0 | 1.0 |
| > 400 to 700 | 1.0 | 1.0 |
| > 700 to 800 | $10\left[\dfrac{\lambda - 700}{515}\right]$ | if: $t \leq \dfrac{10100}{\lambda - 699}$   then: $k_2 = 1.0$    if: $\dfrac{10100}{\lambda - 699} < t \leq 10^4$   then: $k_2 = \dfrac{t(\lambda - 699)}{10100}$    if: $t > 10^4$   then: $k_2 = \dfrac{\lambda - 699}{1.01}$ |
| > 800 to 1060 | $10\left[\dfrac{\lambda - 700}{515}\right]$ | if: $t \leq 100$   then: $k_2 = 1.0$    if: $100 < t \leq 10^4$   then: $k_2 = \dfrac{t}{100}$    if: $t > 10^4$   then: $k_2 = 100$ |
| > 1060 to 1400 | 5.0 | |
| > 1400 to 1535 | 1.0 | 1.0 |
| > 1535 to 1545 | $t \leq 10^{-7}$   $k_1 = 100.0$ <br> $t > 10^{-7}$   $k_1 = 1.0$ | 1.0 |
| > 1545 to 13000 | 1.0 | 1.0 |

Note: The variables in the expressions are the magnitudes of the sampling interval ($t$), in units of seconds, and the wavelength ($\lambda$), in units of nanometers. (From *Federal Register*, Vol. 40, No. 148, Thursday, July 31, 1975)

160

**TABLE 6-7** *Selected numerical solutions for $k_1$ and $k_2$*

| Wavelength (nanometers) | $k_1$ | $k_2$ | | | | |
|---|---|---|---|---|---|---|
| | | $t \leq 100$ | $t = 300$ | $t = 1000$ | $t = 3000$ | $t \geq 10{,}000$ |
| 250 | 1.0 | | | | | |
| 300 | 1.0 | | | | | |
| 302 | 1.0 | | | | | |
| 303 | 1.32 | | | | | |
| 304 | 2.09 | | | | | |
| 305 | 3.31 | | | | | |
| 306 | 5.25 | | | | | |
| 307 | 8.32 | | | | | |
| 308 | 13.2 | | | | | |
| 309 | 20.9 | | | | | |
| 310 | 33.1 | | | 1.0 | | |
| 311 | 52.5 | | | | | |
| 312 | 83.2 | | | | | |
| 313 | 132.0 | | | | | |
| 314 | 209.0 | | | | | |
| 315 | 330.0 | | | | | |
| 400 | 330.0 | | | | | |
| 401 | 1.0 | | | | | |
| 500 | 1.0 | | | | | |
| 600 | 1.0 | | | | | |
| 700 | 1.0 | | | | | |
| 710 | 1.05 | 1 | 1 | 1.1 | 3.3 | 11.0 |
| 720 | 1.09 | 1 | 1 | 2.1 | 6.3 | 21.0 |
| 730 | 1.14 | 1 | 1 | 3.1 | 9.3 | 31.0 |
| 740 | 1.20 | 1 | 1.2 | 4.1 | 12.0 | 41.0 |
| 750 | 1.25 | 1 | 1.5 | 5.0 | 15.0 | 50.0 |
| 760 | 1.31 | 1 | 1.8 | 6.0 | 18.0 | 60.0 |
| 770 | 1.37 | 1 | 2.1 | 7.0 | 21.0 | 70.0 |
| 780 | 1.43 | 1 | 2.4 | 8.0 | 24.0 | 80.0 |
| 790 | 1.50 | 1 | 2.7 | 9.0 | 27.0 | 90.0 |
| 800 | 1.56 | 1 | 3.0 | 10.0 | 30.0 | 100.0 |
| 850 | 1.95 | 1 | 3.0 | 10.0 | 30.0 | 100.0 |
| 900 | 2.44 | 1 | 3.0 | 10.0 | 30.0 | 100.0 |
| 950 | 3.05 | 1 | 3.0 | 10.0 | 30.0 | 100.0 |
| 1000 | 3.82 | 1 | 3.0 | 10.0 | 30.0 | 100.0 |
| 1050 | 4.78 | 1 | 3.0 | 10.0 | 30.0 | 100.0 |
| 1060 | 5.00 | 1 | 3.0 | 10.0 | 30.0 | 100.0 |
| 1100 | 5.00 | 1 | 3.0 | 10.0 | 30.0 | 100.0 |
| 1400 | 5.00 | 1 | 3.0 | 10.0 | 30.0 | 100.0 |
| 1500 | 1.0 | | | | | |
| 1540 | 100.0* | | | 1.0 | | |
| 1600 | 1.0 | | | | | |
| 13000 | 1.0 | | | | | |

*The factor $k_1 = 100.0$ when $t \leq 10^{-7}$, and $k_1 = 1.0$ when $t > 10^{-7}$

Note: The variable ($t$) is the magnitude of the sampling interval in units of seconds.

(From *Federal Register*, Vol. 40, No. 148, Thursday, July 31, 1975)

**TABLE 6-8**  *Accessible emission limits for collateral radiation from laser products*

Accessible emission limits for collateral radiation having wavelengths greater than 250 nm but less than or equal to 13,000 nm are identical to the accessible emission limits of Class I laser radiation, as determined from Tables 6-3 and 6-6 for the appropriate ranges of wavelength and emission duration.

Accessible emission limit for collateral radiation within the X-ray range of wavelengths is 0.5 milliroentgen an hour, averaged over a cross section parallel to the external surface of the product, having an area of 10 square centimeters with no dimension greater than 5 centimeters.

(From *Federal Register*, Vol. 40, No. 148, July 31, 1975)

# 7

# LADDERS, TOWERS, AND
# OTHER AERIAL HAZARDS

Probably in every technician's and engineer's career there is some time when he will go aloft to install or check an antenna. And, in recent years, many CATV technicians have had to climb poles to install and service CATV cable and amplifiers.

Whatever the reason for working above ground level, you should be aware of the following hazards: shock, falling, tower or pole collapse, RF burns, radiation, lightning, rotating antennas, and weather exposure. This chapter discusses these problems and their solutions as related to ladders, towers, and poles.

Safety equipment is a must if you climb more than twenty feet above ground. As a minimum you should wear a safety strap (see Figures 7-1 through 7-3). Much better, however, is the new safe climbing or fall prevention devices (Figure 7-4). Under OSHA regulations, any fixed ladder (as on a tower) longer than twenty feet must be protected by a cage or a safe climbing device. (If a cage is used, a rest platform must be provided every thirty feet.)

The safe climbing devices keep you from falling even if you lose your grip or your footing, or both. Here's how a typical one works: Meyer's Tulito ® uses a steel cable suspended between ladder uprights, from the ground to the top, as a guideline for a special safety belt. (The cable may be added to existing towers or poles.) The climber hooks his safety belt to a special clamp attached to the cable. The clamp slides along the cable as the climber ascends or descends the tower. If the climber starts to fall, a grabbing arm in the clamp almost instantly locks on to the cable, preventing any free fall. A bracket on the climber's safety belt keeps the climber from being turned upside down should he become unconscious.

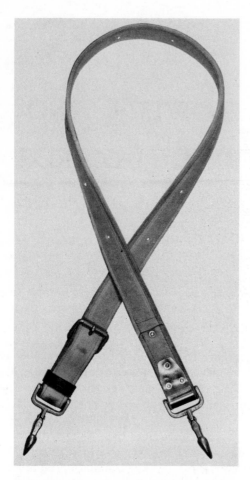

**Figure 7-1.** Safety strap. (Courtesy of W. M. Bashlin Company.)

To keep antennas, towers, and poles from falling, make sure that bases and guy points are securely attached. If they are attached to a building, use screws, etc., instead of nails and attach them to rafters and studs, not sheathing.

Shock dangers aloft fall into three categories: accidental contact of antenna or guy wires with power lines during erection or dismantling, accidental personnel contact with power lines while working aloft, and energized antennas.

It's almost commonplace to read about men being electrocuted while erecting TV, ham, and CB antennas. The causes are usually quite obvious: while raising the antenna, someone overlooked nearby power lines, or overestimated the clearance of the antenna and guys from the power lines, or assumed that if he raised the antenna from a different direction he would

**Figure 7-2.** Lineman's belt.(Courtesy of W. M. Bashlin Company.)

**Figure 7-3.** Lineman's tool holster. (Courtesy of W. M. Bashlin Company.)

**Figure 7-4.** Safe climbing device allows "no-hands" hold. (Courtesy of Meyer Industries.)

not have to worry about clearances (only to have the antenna fall in the "wrong" direction).

To prevent such disasters, all that's necessary is to locate the antenna, guy wires, and lead-in so that if any of them were to break or slip out of your hands it would not make contact with a power line (or get within ten feet of high voltage lines). Since you can easily overestimate clearances, measure them on the ground before you erect the antenna. Do not locate the antenna, guys, and lead-in where a fallen power line could land on them.

For lightning protection, refer to Chapter 4.

To avoid RF burns and exposure to radiation, disable transmitters before going aloft, if possible. And where feasible, disconnect the antenna lead-in even on receivers. A home TV can give you a shock through the antenna because of leakage current which may have originated through shorted capacitors, etc.

To avoid danger from rotating antennas, Navy manual NAVMAT P5100 advises:

> Radar and other antennas which rotate or swing through horizontal or vertical arcs may cause men working aloft to fall. Therefore, the motor switches which control the motion of these antennas should be secured or locked in the open position and suitably tagged before men are permitted to ascend or go within reach of the antenna.

To avoid weather exposure problems, remember that winds aloft may be stronger than those on the ground. In cold weather the chill factor (see Table 7-1) at the top of a tower may be much worse than it is on the ground.

**TABLE 7-1** *Wind chill factor comparisons*

| | Wind Velocity | | | |
|---|---|---|---|---|
| Temperature | Calm | 15 mph | 30 mph | 40 mph* |
| 30 | 30 | 11 | −2 | −4 |
| 20 | 20 | −6 | −18 | −22 |
| 10 | 10 | −18 | −33 | −36 |
| 0 | 0 | −33 | −49 | −54 |
| −10 | −10 | −45 | −63 | −69 |
| −20 | −20 | −60 | −78 | −87 |
| −30 | −30 | −70 | −94 | −101 |
| −40 | −40 | −85 | −109 | −116 |

*Wind speed greater than 40 mph has little additional chilling effect.
(Courtesy of Air Force Inspection and Safety Center.)

## LADDERS

Thousands of people each year are injured, some fatally, by falls from conventional or makeshift ladders, even at heights of less than ten feet. Such falls are caused both by poor climbing practices and by faulty ladders. Safe use of ladders is not as obvious as it may seem—it requires purchase of the proper ladders, careful setup of the ladders, proper climbing techniques, and careful maintenance and storage of the ladders.

*Purchase.* Whether to buy a step ladder, a rolling ladder, or an extension ladder is usually a simple question of height and convenience. The obvious stability of the step ladder and its self-supporting ability are real bonuses in places where there is no support for an extension ladder to rest upon. Wherever possible, buy a ladder that has a seal indicating that it conforms to the standards of the American National Standards Institute (ANSI) or the Underwriters Laboratories (UL).

Aluminum ladders naturally are lighter than wooden ones, but they are deadly around electric power lines since aluminum is such a good conductor.

Even away from power lines, they can be shock hazards if you're using a defective power tool while standing on the ladder. (If you must use power tools while on metal ladders, use double-insulated tools.) Because of the shock hazard, consider purchasing fiberglass ladders, which are nonconductive, if you can afford them. Note that fiberglass and aluminum can stand up to sunlight, moisture, and termites that can ruin wooden ladders left outdoors—they're not going to rot or splinter.

If possible, try out the ladder before purchase. Some have more strength and rigidity than others and will give you a feeling of security because they do not bounce and sway. Make sure you buy a type that is strong enough for your weight and the weight of any objects you may be handling while on the ladder. Look out for sharp edges or burrs on metal ladders.

A light-duty household ladder is not designed for heavy industrial use, obviously. Avoid at all times home-built or job-built ladders (common on many construction sites); their potential faults are too numerous to mention. Look for safety features and accessories before you purchase the ladder: nonskid ladder shoes, pole grip (ladder lash), cable hook, a ladder stabilizer (to straddle windows), a ladder leveler for use where ground surface is not level, rubber or plastic strips for the tops of extension ladder side rails to keep the ladder from slipping on the surface it leans against, and metal ladder brackets to hook over the ridge of the roof.

Buy a ladder that is long enough for any expected uses; too many people try to work from top rungs, which is quite dangerous, because their ladders are too short. Note that the maximum working length of an extension ladder is not as great as its total length because of the overlap of sections, the need for the ladder to extend above surface you will be climbing to, and an allowance for the ladder to stand at an angle. The following table, prepared by the General Services Administration, gives recommended lengths of extension ladders for different heights:

| Height you want to reach | Recommended length |
|---|---|
| $9\frac{1}{2}$ ft | 16 ft |
| $13\frac{1}{2}$ ft | 20 ft |
| $17\frac{1}{2}$ ft | 24 ft |
| $21\frac{1}{2}$ ft | 28 ft |
| $24\frac{1}{2}$ ft | 32 ft |
| 29 ft | 36 ft |

*Setup.* Before using a ladder, it's necessary to follow these safety precautions:

1. Inspect the ladder prior to each use for obvious hazards. On rope-equipped extension ladders, check the rope and the pulley for wear.
2. If it's a very windy day or if a thunderstorm is brewing, avoid ladder work, if at all possible.

3. In carrying the ladder to the work site, keep it horizontal if possible. If carried vertically, it may hit overhead power lines; this can be dangerous even if the lines supposedly are insulated and even if you're using a dry wooden ladder. (Under the worst conditions—uninsulated power line and aluminum ladder—you probably will be electrocuted if you make such contact.)

4. To avoid strain, carry the ladder at its center of gravity (balance point).

5. Set extension ladders the proper distance from the wall (see Figure 7-5). If the base of the ladder is too far out, the ladder is subject to

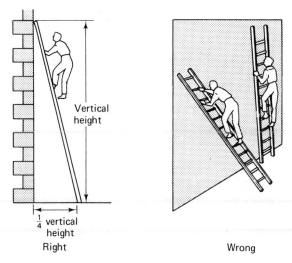

Vertical height

¼ vertical height

Right                                    Wrong

**Figure 7-5.** Proper angle for ladders. (Courtesy of General Services Administration.)

strain that can make it break or slip. However, if the base is too close to the wall, the ladder is likely to tip backward. The experts agree that the proper angle is 75.5 degrees, which means that if you set the ladder base one-fourth of the vertical height from the wall, you will have the safest setup. For example, for a vertical height of twelve feet, the base of the ladder should be three feet from the wall. (The author recognizes the validity of the experts' rule but nevertheless feels unsafe at such a steep angle and moves *his* ladder slightly farther out.)

6. Make sure that your ladder has a firm footing. And keep the feet of the ladder, whether stepladder or extension, level by the use of ladder levelers, brick, concrete blocks, or strong pieces of wood (see Figure 7-6). If the surface is especially smooth or slippery, tie the base of

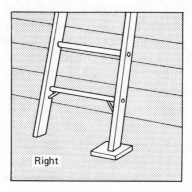

Figure 7-6. Secure footing is essential. (Courtesy of General Services Administration.)

the ladder to an unmovable object, brace it against a fixed object, or have an assistant hold the base of the ladder to prevent slipping.

7. If a ladder must be erected in a walkway or in the path of opening or closing doors, take appropriate steps to protect passers-by and the man on the ladder.

8. Where necessary, use an appropriate ladder device to keep the top of the ladder from moving sideways. Until the top has been secured, have an assistant hold the ladder to keep it from moving sideways.

9. In raising the ladder into position, make sure that electric power lines are not in the way, even if the ladder falls. Get help on long ladders. On extension ladders, make sure the sections overlap sufficiently (three or four ft minimum). Extend the ladder to three ft above the highest landing (roof level) if you intend to work on the roof (see Figure 7-7.) This distance allows you to hold the ladder while stepping on to or off the roof.

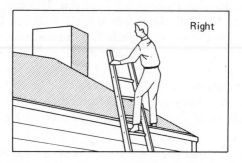

Figure 7-7. Ladder should extend above roof. (Courtesy of General Services Administration.)

10. For fixed ladders, such as towers, keep the lower seven feet protected from unauthorized climbing (e.g., by children). A fence may be placed around the ladder or a fine-mesh wire fence may be wrapped around the ladder itself to prevent climbing.
11. Keep landings at top and bottom of ladders clear and clean.
12. Where possible on long ladders, walk them up into position. Improper lifting can quickly cause back strain.
13. When raising extension ladders, keep your hands on the rails, not the rungs.

*Climbing*
1. Always face a ladder when climbing or descending.
2. Keep your hands free so that they can be used to hold the ladder. Pull up tools and materials with a rope.
3. Make sure your shoes are not wet or muddy.
4. If you wear bifocals, be especially careful of each step.
5. Do not stand on the top two steps of the ladder.
6. Do not leave tools on the ladder.
7. Do not overreach from a ladder or lean too far out—move the ladder closer to the work (see Figure 7-8).
8. Do not step from one ladder to another.
9. Don't use wet ladders around electric circuits.
10. Don't use the ladder as a scaffold.
11. If the ladder is properly secured, you may use a safety belt and strap while you're on the ladder.

**Figure 7-8.** Standing on top of ladder is dangerous. (Courtesy of General Services Administration.)

12. If the ladder is particularly long or unstable, either tie it in place or have someone hold it for you.

13. If a ladder must be left standing unattended for long periods of time, anchor it securely at both top and bottom.

14. Allow only one person on the ladder at a time.

*Maintenance*

1. Do not paint wooden ladders, since this would make it impossible for you to check their condition. Instead, protect them with a coat or two of a clear sealer, spar varnish, shellac, or a clear wood preservative. Use linseed oil to rustproof metal parts.

2. On extension ladders, replace worn or frayed ropes at once.

3. Check any ladder that has been dropped or that has fallen.

4. Do not repair ladders unless you have the necessary experience or guidance to make the repair safely.

*Storage.* Whenever possible, store ladders indoors, in a well-ventilated location, away from dampness and excessive heat.

By necessity, most long ladders must be stored horizontally. If ladder is stored on hangers, use enough hangers (minimum of 3) to prevent sag and permanent set.

## TOWERS

*Erection*

1. Before purchasing a tower, check local, state, and federal laws and building codes for restrictions, safety requirements, etc.

2. Read and follow the tower manufacturer's installation instructions. Only small towers should be erected by the average technician or ham. Specialists are needed to erect large towers because of the tremendous weights and tensions involved. Even for a specialist it's a risky business—in the past year two TV broadcast towers have collapsed and men were killed in each case.

3. A concrete base is recommended even for the smallest tower and of course is essential for the larger towers. Typical dimensions for a 50-ft tower are: the concrete base should be three feet deep and thirty inches square with gravel underneath the base. Follow the manufactorer's instructions. Free-standing towers require larger bases.

4. For concrete base and guy points, let concrete "cure" as long as possible before using. The absolute minimum time should be that recommended by your local concrete contractor or manufacturer.

5. Do a dry run on tower erection so that the actual erection will be performed smoothly and safely.

6. When raising the tower, be prepared for sudden gusts of wind. Have supports on all guys. Keep spectators out of fall range.

7. Make sure that the tower is not so close to power lines that if it fell it would contact the power lines.

8. Use adequate guys and guy wire anchors. If the tower is in a populated area, make sure that the guys have protective shields or guards for the first seven feet above ground to protect children and others from cuts and bruises from running into the guys.

9. If a tower is secured to a building, make sure the bracket is secured to a stud, not just to the siding.

10. Check the tower with carpenter's level to ensure that it is vertical (perpendicular). Survey transits may be used for the same check on tall towers.

11. Apply ample protective coatings to antenna connections so as to minimize future tower climbing or lowering.

12. Use an anticlimb section on the bottom seven feet of the tower to keep unauthorized personnel (especially children) from climbing. You may be able to make your own anticlimb section by wrapping chicken-wire around the legs of the tower. A step-ladder may then be used to get over this section.

13. Tilt-over towers: do not stand under the end when the tower is being tilted.

14. Crank-up towers: Tower sections that get stuck when raising or lowering such towers are a major problem. Therefore, keep hands and feet out of tower, as the sections may "nestle" before you intend them to. Also, be on guard for runaway cranks.

*Maintenance.* Aside from towers erected by hams, inspection and maintenance of other towers should be done only by experienced personnel. Faulty maintenance of large towers can result in disaster. Checking guys for equal tension requires a strand dynamometer and therefore should be left to the experts. However, you don't have to be an expert to check for corrosion, loose or missing bolts, cracks in the tower base, anchor movement, loose or broken connections, or corroded guy wires. Tower inspection should be on a regular schedule to keep minor problems from becoming major. If inspection reveals trouble, get expert help to repair.

Wherever possible, obtain work platforms and other tower accessories from the manufacturer to make antenna maintenance easier and safer.

**Figure 7-9.** Tower work requires safe procedures. (Courtesy of National Cable Television Association.)

*Climbing*
1. Use safety belt when on tower or pole (see Figure 7-9).
2. If possible, do not climb the tower if you must face the sun.
3. Consider hiring a local electrical contractor to use his bucket lift (instead of climbing the tower yourself) if your tower is not too tall and if antenna work is infrequent.
4. Do not climb a crank-up tower unless it is fully lowered or in a nested position.

*Broadcast Management/Engineering Magazine* (January 1967) gives these twelve points for tower safety:

1. Do not climb when even slightly upset or when taking medications such as antihistamines, barbiturates, etc.
2. Always wear the lightest possible clothing; in cold weather provide warm but light-weight jackets, etc.
3. Wear hard hats and suitable boots with soles that do not become slippery when wet.

4. Climbing too fast causes fatigue. Climb at a safe speed. To keep hands warm in cold weather, wear gloves and other handwarming devices.

5. Do not climb alone or at least not without someone near the tower, in the case of smaller towers. (Air Force manual AFM 127-101 states: "Safety observers will be required at the base of antenna poles or towers when climbers are required to ascend higher than 20 ft.")

6. Do not climb without suitable safety belts; make sure they are in good condition (see step 12 under Pole Climbing).

7. Do not climb when thunderstorms are threatening; even when storms seem far away, lightning can be extremely dangerous. Obviously, climbs should not be made when the tower is coated with ice or there is danger that ice may form.

8. Use some means of carrying necessary tools—a bag or pouch that will not interfere with normal arm and leg movements.

9. Maintain communications with workmen on the tower and the ground.

10. Do not service lighting circuits with ac power on.

11. Do not work on transmission lines and other RF-energized elements until the transmitter has been shut off.

12. Practice thorough maintenance of all safety devices and equipment. Lives depend on it.

*Lowering.* Generally, lowering a tower is the reverse of erection. Just remember that a tower can come down too fast if inadequate supports are provided or if supports break.

## POLE CLIMBING

Pole climbing (Figure 7-10) is still a necessary skill in some parts of the country and in certain areas of electronics, even though aerial lift (bucket lift) trucks and underground cables are becoming more and more common. (Aerial lift trucks, it should be noted, cannot always be maneuvered into satisfactory working positions.)

Without proper training and care, you can injure yourself severely during pole climbing. Some of the hazards may be obvious, others not. For example, a pole may be rotted below the ground line, or wasps may have built a nest at the top of the pole. To avoid such problems and to avoid cut-outs (sliding down the pole), follow these rules:

1. Try to obtain formal training in pole climbing.

Figure 7-10. Pole climbing is still a necessity for CATV technicians. (Courtesy of National Cable Television Association.)

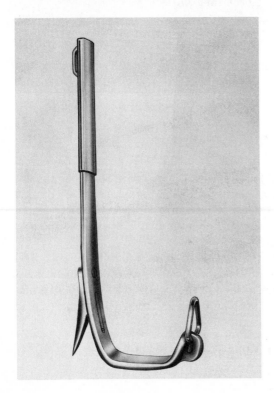

Figure 7-11. Adjustable climbers with replaceable gaffs. (Courtesy of W. M. Bashlin Company.)

2. Do not climb a pole by means of climbers (Figure 7-11) for the first time except under the direction of an experienced climber.

3. Make sure climbers are properly fitted.

4. Inspect climbers daily for nicked or dulled cutting edges on the gaff. Refer to National Safety Council Data Sheet 620 for weekly inspection routine.

5. Wear climbers only when ascending, working aloft, or descending pole.

6. Exercise to loosen joints before climbing.

7. If another man is climbing above you, wait until he is in position before you start climbing. If you're the man on top, don't change your position on the pole until you determine that the man below is in the clear. Whatever you do, don't damage the safety strap of the man below you.

8. When you attach the safety strap to the D-ring, do not rely on the "click" of the snap hook. Look at it to make sure it is properly attached. Do not attach anything to the D-ring except the safety strap.

9. When climbing or descending a pole with climbers, look out for cracks, knots, metal number plates, nails, splinters, shell-rot, etc., any of which could cause a gaff to cut out or otherwise injure you.

10. When taking the last step from the pole to the ground, place your foot down slowly and evenly until you regain your balance.

11. Do not use crossarms, extensions, fixtures, etc., for support until you have first tested them.

12. Take proper care of belts and straps. Don't store them where they may be subjected to excessive heat or dampness. Keep in a cool, dry, well-ventilated area and avoid sunlight during storage. Inspect them frequently. In use, keep the strap flat against the surface without twists or turns.

13. To keep the snap on the strap from coming loose from the D-ring, avoid pressure on the snap-tongue when leaning against arms, pins, braces, wires, etc., which might cause the snap to open.

14. Use special care in checking pine poles, as they deteriorate rapidly.

15. Do not depend on the insulating value of a wood pole.

16. Eliminate wasps with insecticide or freezing aerosol before starting work.

17. It's easy to underestimate wire clearance. If there is any question, use an approved device for measuring the clearance.

18. Men on the ground and linemen must wear approved hard hats. Insulated hard hats are available for linemen. Linemen should have

rubber gloves, protected by outer leather gloves, capable of withstanding 15,000 volts.

## ANTENNA INSTALLATION ON ROOFS

1. Do as much antenna assembly and preparation as possible on the ground.
2. Use proper tools to save time and thereby cut down hazard time—the time you are above ground.
3. Wear nonslipping shoes.
4. On steep roofs, wear safety rope.
5. Use a carpenter's level to ensure that the antenna mast is vertical. (If it's not vertical, the mast will receive unnecessary torque, which may weaken it and cause it to fall.)

# 8

# UNDERGROUND HAZARDS

Throughout the nation and especially in the cities, there is a definite trend toward placing electric power, telephone, and CATV cables underground. Such placement obviously improves the appearance of a community and protects the cables from ice, snow, rain, and wind. At the same time this procedure produces four problems for the electronics technician who is involved in CATV line laying and maintenance:

1. Accidental digups of gas, electric, and telephone lines can kill workers as well as innocent bystanders. Even if no injuries are involved, such digups may inconvenience a lot of people by depriving them of essential utilities.
2. Toxic fumes may be present in manholes where it's necessary to work on cables and amplifiers.
3. In cable-burying operations (Figure 8-1), workers, equipment, materials, and ditches must be protected from oncoming automobiles and trucks and from children, adults, pets, and farm animals. Conversely, the public must be protected from the cable-laying workers and equipment.
4. Deep ditches may cave in and bury alive any worker in the ditch.

The solutions to these problems are straightforward, however cumbersome or difficult they may be to implement.

To avoid digups when laying buried CATV cable, *call before you dig.* Call the local city water and sewer department, the electric department, the gas company, and the telephone company not only for location of pipes and

**Figure 8-1.** Trenches are necessary for some underground CATV cables. (Courtesy of National Cable Television Association.)

cables but for depth as well. While you're at it, check with city hall for performance bonds, hours of work that may be done, requirements for backfilling, and requirements for appropriate barricades and warning devices.

Be prepared for emergencies in case utility-line maps or information are not completely accurate. In California, for example, it has been estimated that only 75% of underground pipelines are on maps of any kind. Obviously the other 25% remain a hazard for the unwary cable layer.

When reliable information is not available, always look for signs which may mark underground cable routes. Inspect a wide area for the signs, not just in your immediate path, as the signs may not be directly over the cables or pipes.

In some cases a metal detector may help you find pipes and cables. But even without such tools, you can always take the simple precaution of hand digging instead of machine digging in suspected danger zones.

If you're fortunate enough to escape injury during a digup, remember that digups can be expensive, as you're likely to be charged for repairing damaged cable or pipe. In south Florida, for example, more than 250 buried telephone cables were accidentally cut in 1972; the average cost to repair each of these cuts was $536, which was billed to the person responsible for cutting the cable.

For these reasons, only qualified personnel should be allowed to operate cable-laying equipment.

**Figure 8-2.** "Buddy" system is safest for work below ground level. (Courtesy of National Cable Television Association.)

To protect yourself from toxic fumes while underground (Figure 8-2), follow the precautions given in Chapter 9.

To protect cable layers from the public and the public from the cable layers, use appropriate guards and warning devices. If equipment must be left unattended, disable it so that inquisitive children and fun-loving college students cannot move it.

Fortunately, most ditches or trenches for CATV cable are not deep; when they are deep, use extensive shoring to prevent cave-ins. When laying cable, keep it a minimum of twelve inches, horizontally and vertically, from power cables.

The National Safety Council (Data Sheet 607) recommends the following procedures in cable laying:

> Whenever foreign objects are encountered during digging, trenching, plowing, or pipe-pushing operations, the operation should be halted immediately and a cautious and thorough investigation made.
> Note: Never cut, chop through, or break off underground obstructions without determining whether they serve a useful purpose. Under no circumstances should underground utility cables, installations, or any pipelines be disturbed.

When underground lines, pipes, or installations of other companies are damaged, the following steps should be taken:

*Electric Lines*

1. Barricade the location until the condition has been cleared.
2. Notify the local power company.
3. Keep the public clear of the area.
4. Notify your supervisor.

*Gas Lines*

1. Leave the hole open to allow the gas to dissipate into the atmosphere. Permit no smoking in the area.
2. Avoid trenching in areas where a spark could set off escaping gas.
3. Warn residents and the public in the vicinity.
4. Notify the local fire department.
5. Notify the local gas company.
6. Keep the public away from the area until the condition is cleared.
7. Notify your supervisor.

*Pipeline other than gas*

1. If a liquid is noticed that appears to be volatile, such as gasoline, follow the same procedure as for gas.
2. If the liquid is not volatile, notify the appropriate utility or municipality.
3. Notify your supervisor.

*Communication Lines*

1. Notify the local telephone company.
2. Notify your supervisor.

### UNDERGROUND COMPACTING AUGER

Contender Corporation's Under-Taker ® underground compacting auger (Figure 8-3) compacts the soil as it bores and saves the trouble and expense of digging up sidewalks and streets when burying cables. Depending on soil conditions, it bores two to seven feet per minute.

Contender Corporation gives these instructions for the auger:

### SAFETY

1. Allow only trained and authorized personnel to operate machine.
2. Do not start engine until power unit and driveline have been inspected, work area is protected, and work area is clear of unauthorized personnel.
3. Do not guide the boring head with anything other than the manufacturer's guide tool properly placed and staked.
4. Do not stand alongside or in front of unit while boring.
5. Do not touch, straddle, or stand alongside of rotating driveline.
6. Remain at the controls while operating unit.
7. Do not operate power cart at right angle to an incline exceeding ten degrees. Operation under such a condition results in a shift of the center of gravity of the power cart, possibly resulting in the cart overturning under high torque operation.

**Figure 8-3.** Under-Taker ® underground compacting auger. (Courtesy of the Contender Corporation.)

8. Stake down driveline every twenty feet to prevent driveline from wandering.

9. Use two trained operators for greatest efficiency and safety.

*PREOPERATION PROCEDURES*

1. Locate all substructures before starting bore. The equipment attempts to penetrate the soil and any obstructions encountered. It cannot distinguish utility lines from other materials, and the avoidance of encounters with and damage to established underground installations is largely dependent on the care and precautions taken by the boring team.

2. Dig a trench at least six inches wide in line with the path the cable is to be buried, beginning at the edge of the area under which the bore is to be made. The length and shape of this starting trench should approximate those indicated in the following diagram. The length of the trench should be about eight to ten times the depth of the bore. At

A — Power cart      D — Guide tool      F — Target trench
B — Segmented driveline   E — Starting trench   G — Surface obstruction
C — Boring head

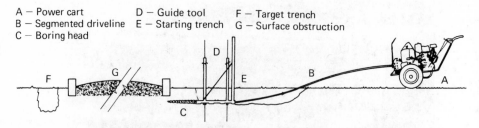

the point where the boring head will enter the soil, the trench should be four to six inches deeper to facilitate proper aiming of the boring head. The boring head and the first section of the driveline should have a straight, unstrained approach to the soil.

3. At the end of the desired path, dig a target trench perpendicular to the path of the bore, approximately four feet long. When the boring head emerges in this trench, it will be necessary to trench along the axis of the bore in order to provide space to uncouple the boring head and attach the appropriate reamer.

4. In certain installations, in order to protect existing substructures such as gas or water lines, it may be necessary to dig pot-holes in the path of the bore to expose these structures to view. When the length of the bore is in excess of seventy-five feet, pot-holes will provide access to the boring head if realignment is necessary. Such openings should be about two feet wide, three feet long and one foot deeper than the projected path of the bore. Distance between pot-holes will depend on total bore distance and the presence of surface and subsurface structures.

*BORING OPERATION*

1. Beginning approximately four feet from where bore is to be started, lay out, end-to-end, several lengths of driveline (not more than sixty feet) along the axis but away from the intended bore.

2. Position the power cart and attach one section of driveline, then couple additional sections in sequence, working toward the starting trench, attaching boring head last.

3. In cold temperatures, start engine and run at low throttle to heat and circulate hydraulic fluid prior to boring. Position the guide tool on the boring head, lower into starting trench and anchor guide tool at desired depth and angle. Careful "aiming" is the key to accurate boring. Always use full throttle for the boring operation. When driveline rotation is started, the firmly anchored guide tool causes the boring head to advance and enter the soil. Boring head screws its way through the soil, drawing the driveline and power cart forward. Stop the rotation

and remove the guide tool after the first length of driveline has entered the soil. Always remain at the power unit controls during operation.

4. Depending on the overall distance of the bore to be made, additional sections of driveline may be necessary. If so, it is recommended that the power cart be stopped when it approaches within twenty feet of the bore. Uncouple the driveline at the power cart and roll the cart back far enough to accommodate the additional driveline.

5. Proceed with the bore until the boring head emerges at a pot-hole or at the target trench.

6. If pot-holes are used, observe the boring head at each pot-hole to determine whether some realignment is necessary. If realignment is necessary, position the guide tool over the boring head and anchor in the desired position to adjust the direction of the boring head.

*LINE-PULLING OPERATION*

1. If the 1.25-inch diameter hole made by the boring head is adequate for the wire or cable to be installed, attach the swivel adapter and swivel to the nose of boring head. When reverse rotation is started, the boring head backs out from target pit to starting trench, bringing with it the cable or conduit. On reverse travel it is necessary to guide the power cart.

2. If the installation requires a hole larger than a 1.25-inch diameter, uncouple the boring head from the driveline at the target pit and install a reamer. Average soils may permit skipping reamer sizes, but dense soils will require the use of each reamer size, in turn, on successive passes through the hole. A swivel is attached to the reamer to install cable or conduit in the hole during the final pass from target pit to starting trench.

## LINE-LAYING MACHINES

The Davis Manufacturing Co. suggests the following precautions in operating a line-laying machine (Figure 8-4):

1. Check to see that the parking brake is set and that the controls are in neutral before starting the engine.

2. Drive at speeds slow enough to ensure safe and complete control, especially when operating on rough terrain.

3. Be particularly careful when going down steep grades or across hillsides.

**Figure 8-4.** Line-laying machine. (Courtesy of Davis Manufacturing Co.)

4. Keep a first-aid kit and fire extinguisher accessible to the operator.

5. Study the terrain on which you are going to operate, and learn the locations and depths of any lines or tubing that might already be buried on the site.

6. Be sure the operator's station is free of oil, ice, or loose objects.

7. Keep the operator's station free of unauthorized persons—no riders.

8. Park on level ground, lower all hydraulic attachments to ground level, stop the engine, and set the parking brake when leaving the machine for lunch or for the night, or at any time the machine is to be left unattended.

9. Become thoroughly familiar with all controls before operating the machine.

10. Operate the machine from the operator's seat; do not start the machine unless you are sitting in the operator's seat.

11. Use seat belts if the machine is equipped with rollover protective structures.

12. Do not smoke or use open flame near the machine while refueling.

13. Do not operate the machine while it is in a closed garage or shed lacking proper ventilation.

14. Do not wear loose-fitting clothes or dangling jewelry that could catch on moving parts. Wear proper safety equipment, such as a hard hat and safety glasses.

In terms of safety, and to avoid plow, cable, or turf damage, take the following steps:

1. Remove the transport lock from the plow.
2. Feed the line into the drop chute before the engine is started. Be sure the line slides freely in the drop chute.
3. Lower the plow blade into the ground a few inches at a time while the machine is moving forward. Failure to do so will result in damage to the drop chute, cable, or both. The blade should be removed from the ground using the same procedure.
4. Coordinate the ground travel speed to soil conditions to avoid tire spin, which could result in turf damage.
5. If the area is particularly rocky or "tight," preplow the area to avoid damage to the cable being buried.
6. Keep the drop-chute swivel linkage free-moving to avoid damage to cable.
7. Be sure skid shoes are properly adjusted to avoid turf damage.
8. Do not stop the forward motion of the machine without stopping the vibration of the line-layer.
9. Be sure lines are free of kinks and coils during the burial operation.

# 9

# TOXIC AND EXPLOSIVE
# CHEMICALS

Whatever their form—gas, liquid, or solid—chemicals can be useful tools in the electronics shop, *or* they can be deadly weapons. It has been estimated that nearly a thousand new chemical compounds are developed each year. In the area of electronics they include cleaning solvents, lubricants, circuit cooling sprays, transformer potting compounds, insulating materials, and gases in waveguides.

Probably most of these chemicals are harmless when proper precautions, usually printed on the container, are followed. However, some chemicals that appear harmless at first may later prove hazardous because of varying environmental conditions. Because of this, new chemicals and new uses for old chemicals must be considered potentially hazardous until proven otherwise.

Toxic and explosive gases may be unintentionally encountered outside the shop or laboratory area when a technician or engineer is on location, for example, installing CATV cables in a manhole.

Just because you have survived exposure to some chemicals over the years does not mean that you can dismiss the danger. Some chemical effects take years for the symptoms to appear in your body. By the time they appear, it may be too late for effective treatment.

The following precautions will protect you from the more common chemical dangers, but it's up to you to be wary of new compounds, and to be alert to newly discovered risks for items that have been in use for years. Carbon tetrachloride, it should be noted, was used for years by thousands of unsuspecting technicians because it was considered safe; now, of course, it's on the forbidden list of chemicals.

## AEROSOLS

Billions of aerosol cans are used each year in hundreds of different applications. Some of the aerosols prepared specifically for use in electronics include TV tuner cleaners and lubricants, insulating sprays, compressed air dusters, corrosion inhibitors, degreasers, circuit coolers, and magnetic tape head cleaners (see Figures 9-1 and 9-2). Most of these are nontoxic *provided they are used as directed.* For your own protection and for that of your coworkers, it's imperative that you heed these instructions:

1. Always read the label on the can. The instructions may have changed since the last container you bought.
2. Keep aerosols away from any source of heat. Under sufficient heat—110 to 120 degrees or more—they may explode with disastrous, possibly fatal consequences. On hot days this means you can't leave them in your car or truck, and even on cooler days you're better off not leaving them in a car trunk. It means also that you must keep them away from direct sunlight, hot water pipes, soldering irons and torches, and incinerators.
3. Do not puncture an aerosol can; there may be more gas left than you realize.
4. Use adequate ventilation, which means opening the window and turning on the exhaust fan, when using aerosols. A closed room with air conditioning may not provide sufficient exhaust, so do not depend

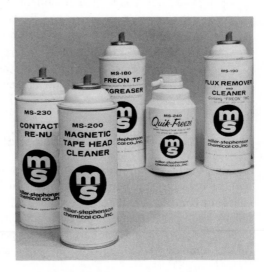

**Figure 9-1.** Typical assortment of aerosols for electronics use. (Courtesy of Miller-Stephenson Chemical Company.)

**Figure 9-2.** Moisture and fungus-proof varnish in use. (Courtesy of Sprayon Corp.)

on it. Consider carrying your work and your aerosol can outdoors if it is practical. Whether inside or out, take a deep breath before spraying and hold it as long as reasonable to avoid inhalation.

5. Aim the aerosol spray accurately. Wherever possible, use extension nozzles to pin point the spray. Avoid the eyes and skin when spraying, particularly when using circuit coolers, which can cause frostbite. Do not spray in or near a flame; while the spray itself may be nonflammable, it may decompose in the flame and form toxic gases from what was a harmless gas.

## BATTERIES

Lead-acid storage batteries have two dangers: the sulfuric acid is highly corrosive and battery gases (primarily hydrogen) are explosive. In charging batteries or jumping a dead one, remove vent caps and keep all flames and sparks away to prevent ignition. Make sure fluid is at proper level before proceeding. On batteries equipped with flame arrester vent caps, do *not* remove the vent caps during jump starting. Make sure the booster battery and the "dead" battery have the same voltage. Connect the positive terminal of the booster battery to the positive terminal of the dead battery (assuming that the dead battery has the negative terminal grounded.) Connect the negative terminal of the booster battery to the chassis of the dead-battery car about twelve inches away from the negative terminal of the dead battery. *Do not connect directly to the negative terminal on the dead battery* because of the possibility of explosion. Turn on the assisting vehicle engine and allow it to run for a few minutes. Then try to start the engine in the car with the dead battery. Reverse the above procedure when removing the jumper cables. Watch the engine fan blade during this procedure.

When working in a battery storage and charging room, make sure that:

1. A deluge shower is nearby for neutralizing acids.
2. Floors are clean and dry.
3. Proper ventilating equipment is installed to prevent accumulation of explosive mixtures. The ventilation system should be interlocked with the battery charging system so that battery charging will cease if the ventilation system fails.
4. No one is smoking.
5. A fire extinguisher is handy.

When removing a battery from a circuit, make sure the power switch is turned off. Otherwise, the spark created when the battery is disconnected may be sufficient to ignite any hydrogen that has accumulated.

When a battery must be discarded, do not incinerate it. Even if an explosion does not occur, toxic fumes may be released during the burning.

Keep heat away from batteries. Accidental shorting may cause enough heat to trigger an explosion, particularly with mercury batteries.

Do not use a multicell dry battery after its closed circuit voltage has dropped below 0.9 volt per cell, as further discharge may generate hydrogen (Navy manual NAVMATP 5100).

*NICKEL–CADMIUM BATTERIES*

*Aerospace Safety Magazine* (November 1974) recommends these procedures when working with nickel–cadmium batteries:

1. Keep lead–acid batteries and tools used with them away from Ni–Cd batteries.
2. Before adjusting the electrolyte, make sure the battery is fully charged (which can be determined only by monitoring the charging input in current and time until the ampere-hour capacity of the battery has been reached). After this is accomplished, allow the battery to set for at least two hours; then adjust the electrolyte level only if it is required.

3. Note that this is opposite from the procedure used with lead–acid batteries.

## CADMIUM

Cadmium is widely used in electroplating, in some solders, and in alkaline batteries. Cadmium-plated screws are commonly specified for use in corrosive atmospheres. When vaporized through heating (such as welding) or burning (through incineration or accidental fires), cadmium can be the source of deadly vapors. In addition to the inhalation danger, there is the problem of the absorption of particles of cadmium through a body wound or cut.

## HAZARDOUS LOCATIONS

At locations where combustible dusts are an explosion hazard, the National Safety Council recommends good housekeeping, dust and ignition-resistant enclosures, sealing, adding inert dusts, purging, and the use of intrinsically safe and nonincendiary equipment to control the hazard. Although this is not a problem in the ordinary shop, it can become one for a technician on location, as, for example, in the repair of industrial electronic equipment in a flour mill.

## LIQUID CRYSTALS

*Popular Electronics*[1] magazine recommends the following procedures when handling nematic liquid crystal material:

> At no time should you allow this material to contact your flesh or clothing, nor should you under any circumstances breathe in its fumes. Do as professional chemists do: wear surgical rubber or polyethylene (throw-away) gloves at all times and have an exhaust fan going to draw away the fumes. Should the liquid spill on a workbench or counter top, immediately clean it up and thoroughly scrub the area with paper towels.

## OZONE

Ozone is the gas with a pungent odor that you sometimes smell after intense thunderstorms or even around electrical equipment that forms arcs. It's a highly active form of oxygen with a vicious reputation; in fact, it's 1,000 times more toxic than carbon monoxide. Electronic air cleaners have been known to produce some ozone, but whether they produce a hazardous

[1]March 1973, Copyright 1973, Vol. 3 No. 3, p. 9, Ziff-Davis Publishing Company.

amount or not is debatable. Until the debate is resolved, the safest policy is to stay away from them; if you can't, be on guard for unusual odors, however faint, you may encounter when you turn such units on. Replace the charcoal filter (if one is used) if you smell ozone. If that fails, have the unit checked for defects.

## PRINTED CIRCUITS

Solder, etchants, and solvents used in production of printed circuit boards create fumes that should be exhausted to the outside. Also, avoid skin contact with these chemicals—wear gloves. Wear goggles when cleaning p.c. boards.

## SUBSURFACE GASES

With the trend to underground placement of power and communications lines, including CATV cable, some technicians must now occasionally work below the ground surface, in manholes and cable vaults.

Unfortunately, subsurface structures or even holes in the ground may seem deceptively innocent and yet harbor highly toxic gases that can kill you before you can signal for help. Even if you avoid asphyxiation or poisoning, you may encounter explosions and fires in these locations. To stay on the safe side (see Figure 9-3):

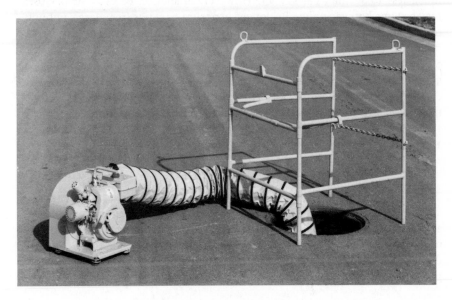

**Figure 9-3.** Portable ventilating blower provides fresh air to manholes. (Courtesy of Over-Lowe Company.)

1. Don't depend on your nose to detect gases. Some poisonous gases have no odor. Others can paralyze your sense of smell.
2. Instead, use appropriate gas detection equipment. Refer to National Safety Council Data Sheet 550 for proper test techniques. Test the atmosphere in the manhole before entering. Check for safe oxygen level. So-called nontoxic gases can kill—when they reduce the oxygen content of the air below the critical level. Under such conditions, nitrogen, helium, and freon have killed people.
3. Use forced-air ventilation even if the area checks out safely.
4. Do not work alone. Have a buddy outside the manhole hold your lifeline and watch for trouble.
5. Keep unauthorized equipment and tools outside the manhole, as sparks could ignite gases present. By the same token, keep all open flames and cigarettes away from the manhole.
6. Keep a fire extinguisher handy.
7. During rain or snow you can use a protective tent over your head, but do not completely close the tent.

## TRICHLOROETHYLENE

Trichloroethylene is often recommended as a substitute for carbon tetrachloride; it does essentially the same cleaning and degreasing job but is much less toxic than carbon tet. However, inhalation of even small doses may become habit-forming. Prolonged exposure or exposure to large amounts of this chemical can give you symptoms of being intoxicated and can wreck your heart and nervous system. The National Institute of Occupational Safety and Health has recommended a 100-ppm (parts per million) limit on an eight-hour time-weighted average. According to some authorities, if you can smell trichloroethylene, you're probably being overexposed. Therefore, make sure you have adequate ventilation when you use it, and wear a face shield if there's any possibility of the chemical being splashed on your face. Keep it away from open flames or high temperature processes.

## WAVEGUIDES

Waveguides for use with moderate-power microwave equipment use dry air as the dielectric medium. With high-power equipment, however, it's necessary to replace the dry air with a gas that has a higher dielectric strength. Air Force Technical Order T. O. 31Z-10-4 provides the following safety guidelines for waveguide gases:

In many applications, microwave transmission lines are quite long and subject to leakage when pressurized. Present standards for waveguides used in ground radar equipment specify that the leakage rate for a 30-psig system must not exceed 1.25 psi per hour. The waveguide originating at a transmitter usually passes through areas accessible to, and possibly occupied by, personnel; therefore, the toxicity factor of the gas used to pressurize the waveguide must be known, to safeguard personnel against toxic effects arising from leaks in the waveguide, especially should the leak occur in closed spaces. Adequate ventilation, provided by exhaust fans, may be required to keep the concentration of escaping gas well below a tolerable percentage in closed spaces occupied by personnel.

The use of either Freon or sulfur hexafluoride as a dielectric medium to pressurize waveguide systems does permit increasing the power handling capability of the waveguide system. However, in the event of arc-over or breakdown, both gases are subject to decomposition. Freon is not likely to be used in low-temperature applications because of its comparatively high condensation temperature. However, it is important to note that after breakdown, one of the decomposition products of Freon is phosgene, which is a highly toxic gas and extremely dangerous to personnel.

Sulfur hexafluoride in its pure state is essentially inert and nontoxic, and has found use in medical applications as a therapeutic measure to rehabilitate damaged lungs. In tests on humans, the gas in its pure state has been found to be nontoxic when inhaled in gas-oxygen mixtures containing as much as 80% sulfur hexafluoride. However, when arc-over occurs in a waveguide filled with this gas, the decomposition products that are produced constitute a dangerous personnel hazard in the form of several toxic gases, including fluorine. These toxic gases may not irritate the skin, are colorless, and cannot be detected by odor, but will cause extreme lung irritation and hemorrhaging.

Where sulfur hexafluoride is used to pressurize waveguide systems:

a. The portion of the waveguide system which originates at the transmitter and passes through confined areas should be well sealed and made as gas-tight as is possible.

b. A room ventilation system should be provided in confined areas where leakage from the waveguide is possible. The room exhaust fan should provide a complete change of air every several minutes to prevent concentration of toxic gases.

c. Consideration should be given to the incorporation of an escape valve in the waveguide system, at a point external to any closed area or equipment shelter, to allow continuous leaking of the gas to the open atmosphere. The purpose of such a pressure leak is to keep the gases moving through the system and to expel decomposition products resulting from electrical breakdown to the open atmosphere, where the gases will be diluted and dissipated harmlessly. The rate of flow through the escape valve would require adjustment to a rate which is economically feasible.

# 10

# NOISE

Noise has been a nuisance, if not a hazard, for centuries. In recent times, though, noise levels have increased to the point where they are a real threat to our health. In fact, noise is probably the single greatest danger in the work environment. As many as 16 million Americans may be working at jobs that have hazardous noise levels. Already 3 million people have lost some or all of their hearing because of noise.

Thus, noise is not simply an annoyance that one must adapt to; the only adaptation possible is to go completely or partially deaf. If noise is not a problem to you, it may be because your hearing has already been damaged.

The average person is almost totally unaware of his vulnerability to hearing loss, according to Dr. Donald Belt of the Stanford Medical Center. This is due, in part, says Belt, to the fact that hearing loss from noise exposure is usually insidious and deceptive. There is neither pain nor bleeding. After noise exposure, the individual may have the feeling of a temporary hearing loss, some ringing sounds in his ears, and a slight decrease in his ability to communicate effectively. Although the symptoms may disappear after a night's rest, the fact is that his hearing is not completely restored, and after many repeated insults an accumulation of effects bring about a significant and irreversible loss of hearing.

To determine if you work in a hazardous noise area, try this simple test (see Figure 10-1): do you have to raise your voice to talk to someone standing beside you? If so, you had better take steps to protect your hearing.

Before starting noise abatement or control, you should be aware of the basic principles of sound measurement.

Sound (or noise), as you probably know, is measured in decibels (dB)

196

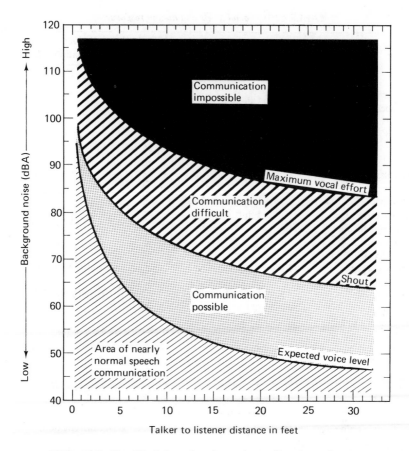

**Figure 10-1.** Simplified chart that shows the quality of speech communication in relation to the *A*-weighted sound level of noise (dBA) and the distance between the talker and the listener. (From *Effects of Noise on People*, NTID300.7, U.S. Environmental Protection Agency.)

on a scale which varies from 0 dB, the threshold of hearing, to 120 dB, the threshold of pain (see Tables 10-1, 10-2, and 10-3). In between these limits, of course, is a wide range of sound levels. The gentle rustle of leaves is rated at 10 dB; daytime residual noise level in a wilderness is about 16 dB; and a typical office has a background noise level of 50 dB. Outdoor daytime noise levels are 30 to 35 dB on farms and 60 to 75 dB in the city. When the outdoor median noise levels are above 71 dB, special soundproofing is necessary to preserve the indoor noise environment.

Strictly speaking, such measurements must be related to the particular scale on which they are read on a sound-level meter. Usually this is the *A* scale, which discriminates (through special circuits) against very low fre-

**TABLE 10-1**  *Sound levels and human response*[1]

| | Noise Level | Response | Hearing Effects | Conversational Relationships |
|---|---|---|---|---|
| | 150 | | | |
| Carrier Deck Jet Operation | 140 | | | |
| | | Painfully Loud | | |
| | 130 | Limit Amplified Speech | | |
| Jet Takeoff (200 feet) | 120 | | ← CONTRIBUTION TO HEARING IMPAIRMENT BEGINS | |
| Discotheque Auto Horn (3 feet) | | Maximum Vocal Effort | | |
| Riveting Machine | 110 | | | |
| Jet Takeoff (2,000 feet) | | | | |
| Garbage Truck | 100 | | | Shouting in ear |
| N.Y. Subway Station Heavy Truck | | Very Annoying Hearing Damage | | Shouting at 2 ft. |
| (50 feet) | 90 | (8 hours) | | |
| Pneumatic Drill (50 feet) | | | | Very loud |
| Alarm Clock | 80 | Annoying | | Conversation, 2 ft. |
| Freight Train (50 feet) | | | | Loud |
| Freeway Traffic | 70 | Telephone Use | | Conversation, 2 ft. |
| (50 feet) | | Difficult Intrusive | | |
| Air Conditioning | 60 | | | Loud Conversation, 4 ft. |
| Unit (20 feet) | | | | |
| Light Auto Traffic | 50 | Quiet | | Normal Conversation, 12 ft. |
| (100 feet) | | | | |
| Living Room Bedroom | 40 | | | |
| Library | | | | |
| Soft Whisper (15 feet) | 30 | Very Quiet | | |
| Broadcasting Studio | 20 | | | |
| | 10 | Just Audible | | |
| | 0 | Threshold of Hearing | | |

[1](Courtesy of the Environmental Protection Agency.)

TABLE 10-2   *Noise in the home*[1]

| Item | Average dB |
|------|:----------:|
| Air conditioner | 55 |
| Alarm clock | 60 |
| Blender-electric | 93 |
| Can opener-electric | 78 |
| Clothes dryer-automatic | 64 |
| Coffee grinder | 69 |
| Conversation-average | 45 |
| Conversation @ 3′ | 64 |
| Dishwasher | 69 |
| Doorbell | 100 |
| Drill-$\frac{1}{4}''$ portable | 70 |
| Fan-12″ portable | 70 |
| Fan-vent | 63 |
| Fan-wall exhaust | 90 |
| Furnace blower | 100 |
| Garbage disposer | 78 |
| Hair dryer | 77 |
| Jig saw | 68 |
| Knife sharpener | 78 |
| Mixer-electric | 82 |
| Pots and pans | 73 |
| Radio | 78 |
| Range vent fan and dishwasher | 86 |
| Range vent fan and disposer | 91 |
| Refrigerator | 45 |
| Sander-belt | 91 |
| Sander-disc | 93 |
| Sander-orbital | 70 |
| Saw-8″ radial | 92 |
| Saw-sabre | 76 |
| Saw-6″ skill | 100 |
| Sewing machine | 64 |
| Shaver-electric | 85 |
| Shower | 78 |
| Sink drain | 86 |
| Telephone ring @ 6′6″ | 78 |
| Toilet-flushing | 67 |
| Toilet & vent fan | 71 |
| TV | 68 |
| Vacuum cleaner | 84 |
| Washing machine-automatic | 64 |
| Water faucet | 68 |
| Whisper @ 5′ | 10 |
| Wood lathe | 80 |

[1](Courtesy of the Koss Corp.)

**TABLE 10-3**[1]   *Order-of-magnitude estimates of exposure to home appliance and building equipment noise expressed in millions of person-hours per week*

| Noise Source | Speech Interference* Moderate | Severe | Sleep Interference* Slight | Moderate | Hearing Damage Risk Slight | Moderate |
|---|---|---|---|---|---|---|
| Group I: Quiet Major Equipment and Appliances | | | | | | |
| Fans | 1200 | | 0 | | 0 | |
| Air Conditioner | 242 | | 121 | | 0 | |
| Clothes Dryer | 94 | | 10 | | 0 | |
| Humidifier | 10 | | 15 | | 0 | |
| Freezer | 0 | | 0 | | 0 | |
| Refrigerator | 0 | | 0 | | 0 | |
| Group II: Quiet Equipment and Small Appliances | | | | | | |
| Plumbing (Faucets, Toilets) | | 535 | 267 | | 0 | |
| Dishwasher | | 461 | 4 | | 0 | |
| Vacuum Cleaner | | 280 | 0.5 | | 0 | |
| Electric Food Mixer | | 222 | 1 | | 0 | |
| Clothes Washer | | 215 | 0.5 | | 0 | |
| Electric Can Opener | | 117 | 0.2 | | 0 | |
| Electric Knife | | 1 | 0.1 | | 0 | |
| Group III: Noisy Small Appliances | | | | | | |
| Sewing Machine | | 19 | 0.5 | | 9 | |
| Electric Shaver | | 6 | 1 | | 5 | |
| Food Blender | | 2 | 0.2 | | 0.5 | |
| Electric Lawn Mower | | 1 | 1 | | 0.3 | |
| Food Disposer | | 0.5 | 0.5 | | 0.5 | |
| Group IV: Noisy Electric Tools | | | | | | |
| Home Shop Tools | | 5 | 2 | | 1 | |
| Electric Yard Care Tools | | 1.5 | 0.1 | | 0.4 | |

*These figures are not directly interpretable in terms of person-hours of lost sleep or speech interference.
[1](Courtesy of the Environmental Protection Agency.)

quencies, just as does the human ear. Thus, a sound-level meter set to the *A* scale will respond to sound in the same manner as the human ear. Measurements on the *A* scale are recorded as dBA, although frequently the *A* may be missing and is assumed. Many professional societies and OSHA have accepted the *A* scale as standard.

In measuring and comparing noises, note that decibels for noise measurement cannot be added or subtracted arithmetically. For example, if a sound level is doubled, the corresponding increase will be 3 dB, not double the number (see Figure 10-2). Consequently, if a 90-dB machine is placed next

76 dB                                                 79 dB

**Figure 10-2.** Doubling the number of identical sources results in a
3-dB increase in sound pressure level. (From *Fundamentals of
Noise: Measurement, Rating, Schemes, and Standards*, NTID300.15,
U.S. Environmental Protection Agency.)

to another 90-dB machine, the total noise level will be 93 dB, not 180 dB.
The correct way to add noises is to change the sound pressure levels to ratios
of sound intensities, add the ratios, and then reconvert to decibels.

## EFFECT OF NOISE ON MAN

At any intensity, noise can be irritating and adversely affect your disposi-
tion, if not your health, by interrupting sleep, interfering with conversation,
or by disturbing relaxation. On the job it may keep you from necessary concentration on a complicated task. The stress produced by noise may con-
tribute to heart disease, gastric ulcers, and high blood pressure, according to
one current theory.

Although some of the effects of noise on man are debatable, there's no
doubt that noise can cause temporary and permanent hearing losses (see
Tables 10-4 and 10-5).

**TABLE 10-4**  *Factors influencing the degree and extent of
noise-induced hearing loss*[1]

Intensity or loudness of the noise level
Duration of noise exposure each day (duty cycle per day)
Temporal distribution of noise exposure
   Continuous noise
   Intermittent steady (equal intensity) noise
   Intermittent variable intensity noise
Frequency distribution (spectral content)
Character of surroundings in which the noise is produced
Distance from source
Position of ears in respect to sound waves
Total work duration (years of employment)
Individual susceptibility
Age of the worker
Co-existing hearing loss and ear disease

[1](Courtesy of National Safety News, April 1973.)

TABLE 10-5   *Hearing handicap guideline*[1]

| Class | Degree of Handicap | Average Hearing Threshold Level for 500, 1000, and 2000 Hz in the Better Ear* | | Ability to Understand Speech |
|-------|--------------------|-----|------|---------------------------|
| | | More Than | Not More Than | |
| A | Not significant | | 25 dB | No significant difficulty with faint speech |
| B | Slight Handicap | 25 dB | 40 dB | Difficulty only with faint speech |
| C | Mild Handicap | 40 dB | 55 dB | Frequent difficulty with normal speech |
| D | Marked Handicap | 55 dB | 70 dB | Frequent difficulty with loud speech |
| E | Severe Handicap | 70 dB | 90 dB | Can understand only shouted or amplified speech |
| F | Extreme Handicap | 90 dB | | Usually cannot understand even amplified speech |

*Measured in a properly designed audiometric examination facility using an audiometer calibrated to meet ANSI standards.
[1](Courtesy of the Environmental Protection Agency.)

Just how does noise affect your ears? The Environmental Protection Agency puts it this way: When sound enters the ear, the waves pass through the ear canal to the eardrum which vibrates. The eardrum conducts these vibrations to three tiny bones called *ossicles*—the three tiniest bones in the body. It is here that the acoustic reflex occurs. The ossicles change the loudness of sound before it enters the inner ear (Figure 10-3). Normal action of the ossicles may amplify soft sounds or dampen loud sounds as their tiny muscles contract to decrease the pressure of the sound waves.

The acoustic reflex protects the inner ear from extra loud sounds by reducing them, just as the eye protects itself from extra bright light by contracting the pupil. However, the ear is not completely successful in this task. The reason is twofold: first, the reflex occurs on command from the brain a few hundredths of a second after the loud sound is first sensed. Thus, at least some of the sound at full loudness gets through to the delicate inner ear before the reflex goes into operation. Second, muscles cannot contract indefinitely, so their sound dampening capacity is limited. Therefore, if the loud sound is sustained, the inner ear may still be bombarded with excessive sound pressure even after the reflex has had a chance to work. In the case of impulse sounds such as a gunshot, the reflex is virtually useless as a defense.

What happens when loud sounds enter the inner ear? The ossicles transmit the vibrations to a fluid contained in a tiny, snail-shaped structure called the *cochlea*. Within the cochlea are microscopic hair cells that move back and forth in response to the sound waves just as seaweed on the ocean floor

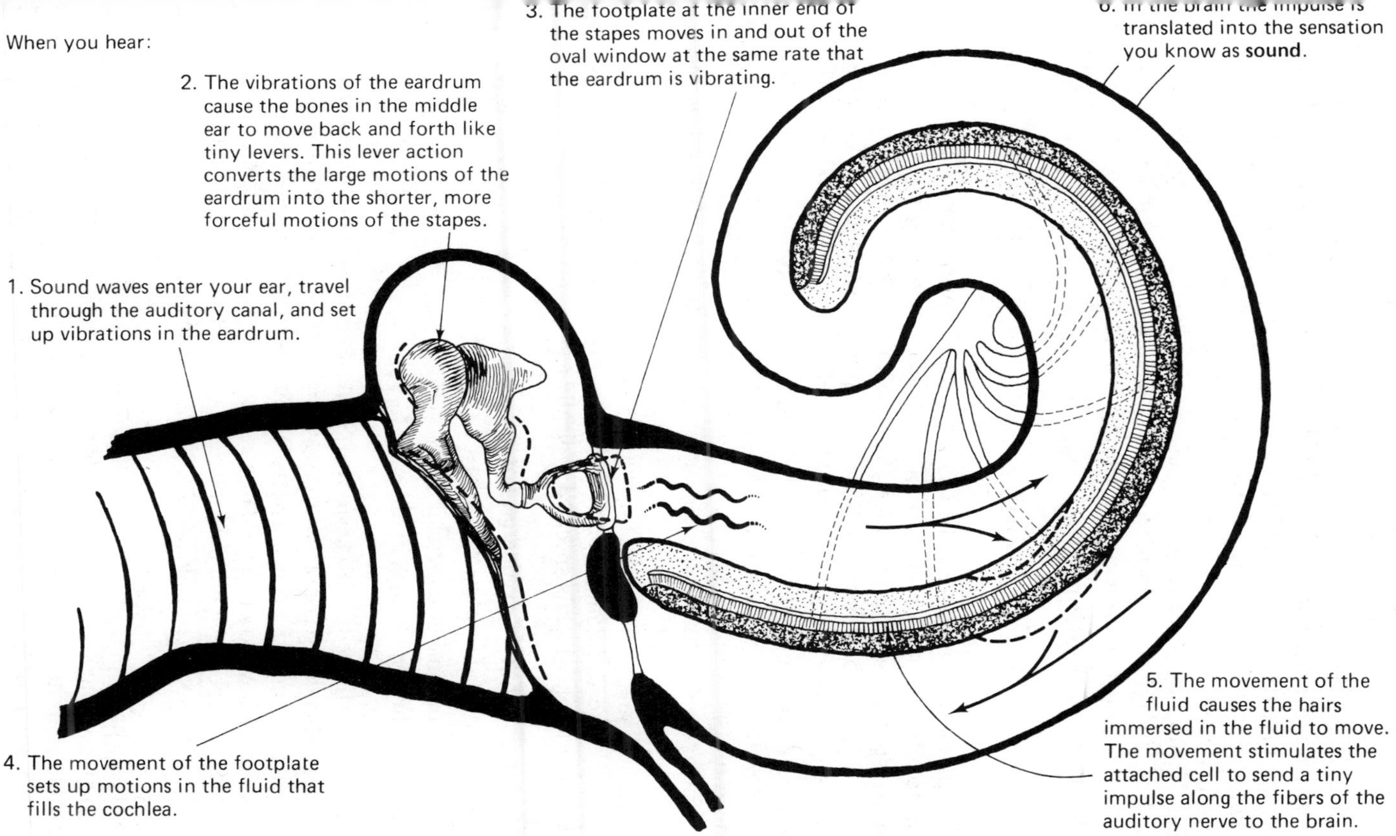

When you hear:

1. Sound waves enter your ear, travel through the auditory canal, and set up vibrations in the eardrum.

2. The vibrations of the eardrum cause the bones in the middle ear to move back and forth like tiny levers. This lever action converts the large motions of the eardrum into the shorter, more forceful motions of the stapes.

3. The footplate at the inner end of the stapes moves in and out of the oval window at the same rate that the eardrum is vibrating.

4. The movement of the footplate sets up motions in the fluid that fills the cochlea.

5. The movement of the fluid causes the hairs immersed in the fluid to move. The movement stimulates the attached cell to send a tiny impulse along the fibers of the auditory nerve to the brain.

6. In the brain the impulse is translated into the sensation you know as **sound**.

**Figure 10-3.** The workings of the ear. (Courtesy of the National Bureau of Standards.)

203

undulates in response to wave action in the ocean. It is the energy impulses created by the movement of these crucial hair cells that go to the brain where they are interpreted as sound. But just as the seaweed can be torn and ripped by violent wave action in the ocean, so too can hair cells be damaged by too intense sound waves.

When intense sound waves occur only briefly, the damage may be temporary. But if loud noises are frequent or sustained, the damage may be permanent, and such noise-induced hearing loss cannot be restored either through surgery or hearing aids.

## NOISE LAWS

Throughout the country, both city and state governments have been enacting tough new noise control laws and have been sending out noise pollution inspectors to enforce these laws. Although there may be great differences in the effectiveness of these laws and the strictness with which they are enforced, the new noise control legislation reflects the public's growing intolerance of noise; it also closes many of the loopholes in previous noise control laws.

Instead of using phrases such as "disturbing the peace" these ordinances specify, for example, that loudspeakers for advertising on public streets or for attracting public attention to a building are forbidden. Or they may specify that permits are required for sound trucks.

Some of these ordinances define noise as any sound that exceeds the ambient noise level, which is specified typically as:

45 dBA for nighttime residential areas
55 dBA for daytime residential areas
65 dBA for anytime commercial areas
70 dBA for anytime all other zones

(The National Institute for Occupational Safety and Health defines *ambient noise* as the all-encompassing noise associated with a given environment, being usually a composite of sounds from many sources near and far.)

Operation of a radio, TV, or stereo so as to be plainly audible at a distance of fifty feet from the building in which it is located during the hours of ____ to ____ is considered to be a *prima facie* evidence of a violation in some cities.

Two basic principles of law are incorporated in most noise ordinances, even though not stated: (1) everyone must put up with a certain amount of aggravation or annoyance (or noise, in this case) and (2) the amount you have to tolerate must be weighed against the benefit or necessity of the noise (or sound) to the noise maker.

As the public becomes more aware of noise and the measures that must be taken to control it, more and more noise ordinances are likely to be passed. Any technician or engineer who operates sound amplifying equipment (particularly the outdoor type) or noisy shop equipment should investigate local ordinances to avoid fines.

By far the most important law relating to noise is the noise-control provisions of the Occupational Safety and Health Act. Essentially, OSHA says that if the sound level exceeds 90 dBA where you work, you must wear hearing protectors or limit the number of hours you work in such noisy environments. The complete text of the act as it pertains to noise is as follows:

Protection against the effects of noise exposure shall be provided when the sound levels exceed those shown in Table 10-6 when measured in the *A* scale of a standard level meter at slow response. When noise levels are determined by octave band analysis, the equivalent *A*-weighted sound level may be determined from Figure 10-4.

Octave band sound pressure levels may be converted to the equivalent *A*-weighted sound level by plotting them on the illustrated graph and noting the *A*-weighted sound level corresponding to the point of highest penetration into the sound level contours. This equivalent *A*-weighted sound level, which may differ from the actual *A*-weighted sound level of the noise, is used to determine exposure limits from Table 10-6.

When employees are subjected to sound exceeding those listed in Table 10-6, feasible administrative or engineering controls shall be utilized. If such controls fail to reduce sound levels within the levels of the table, per-

**TABLE 10-6**  *Occupational safety and health act*

| Duration per day, hours | Sound level Slow response dBA |
|---|---|
| 8 | 90 |
| 6 | 92 |
| 4 | 95 |
| 3 | 97 |
| 2 | 100 |
| $1\frac{1}{2}$ | 102 |
| 1 | 105 |
| $\frac{1}{2}$ | 110 |
| $\frac{1}{4}$ or less | 115 |

Permissible Noise Exposure[1]

[1]When the daily noise exposure is composed of two or more periods of noise exposure of different levels, their combined effect should be considered, rather than the individual effect of each. If the sum of the following fractions: $C1/T1 + C2/T2, \ldots, Cn/Tn$ exceeds unity, then, the mixed exposure should be considered to exceed the limit value. *Cn* indicates the total time of exposure at a specified noise level, and *Tn* indicates the total time of exposure permitted at that level.

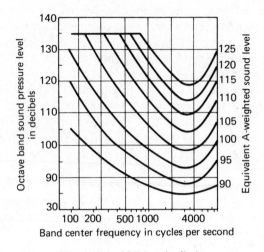

**Figure 10-4.** OSHA noise limits.

sonal protective equipment shall be provided and used to reduce sound levels within the levels of the table.

If the variations in noise levels involve maxima at intervals of one second or less, it is to be considered continuous.

In all cases where the sound levels exceed the values shown herein, a continuing, effective hearing conservation program shall be administered.

Exposure to impulsive or impact noise should not exceed 140 dB peak sound pressure level fast response.

By *administrative controls* OSHA means that workers in extremely noisy jobs should be moved away from the noise after they have used up their exposure limits as indicated in Table 10-6. *Engineering controls* means the use of various types of soundproofing and noise conditioning materials and devices.

Some feel that OSHA's standards are not strict enough; the National Institute for Occupational Safety and Health (NIOSH) has recommended to OSHA that OSHA's occupational noise exposure level of 90 dBA for an eight-hour day be lowered to 85 dBA. Other recommendations by NIOSH include:

1. A warning sign shall be appropriately located at the entrances to and/or at the periphery of areas where there exists sustained environmental noise at or in excess of the limits prescribed. The notice shall read: *Warning, Noise Area, May Cause Hearing Loss, Use Proper Ear Protection.*

2. The use of protective equipment to prevent occupational noise expo-

sure of the employee in excess of the prescribed limits is authorized only until engineering and administrative controls and procedures can be implemented to maintain the occupational noise exposures within prescribed limits.

3. Each worker exposed to noise shall be apprised of all hazards, relevant symptoms, and proper conditions and precautions for working in noisy areas.

4. Employers will be required to maintain records of (a) environmental exposure monitoring for a period of ten years, (b) all audiograms for a period of twenty years, and (c) all audiometric calibration data for a period of twenty years.

NIOSH states that approximately 14% of workers in manufacturing are exposed to noise above 90 dBA, but it is not known how many are exposed to more than 85 dBA.

NIOSH's recommendations may not be incorporated into OSHA's regulations for several more years because of the expense of installing sufficient soundproofing to hold noise levels down to 85 dBA. Manufacturers are already complaining that the 85-dBA level would cost them millions of dollars. Nevertheless, 85 dBA is a level that should be sought in the construction of all new workshops, laboratories, offices, and factories.

## NOISE MEASUREMENT

In our noise-polluted world, it's possible to make an intelligent guess of the loudness of a particular noise without using an instrument. However, to satisfy government laws and to make any reasonable attempts at noise abatement, it's necessary to measure the noise with sound level meters (Figure 10-5), noise analyzers, or dosimeters (Figure 10-6).

The basic, most common instrument is the sound level meter ($75 to $340), which consists of a microphone, audio amplifier, and meter, all packaged for portable use. Generally it contains one or more weighting networks to ensure that the meter responds to sound in the same way as the human ear. That is, *loudness* is determined by pitch as well as sound pressure level. The most common and useful weighting network that achieves this effect is the *A*-scale.

Although not always essential, an accessory calibrator (around $300) allows the user to ensure the accuracy of his sound level meter.

Noise dosimeters (about $1,000) give a cumulative measurement for an individual who may move in and out of high noise areas during his workday. Under OSHA noise regulations, for all people who work on jobs that have noise levels greater than 90 dBA, it's necessary to keep track of the employee's

**Figure 10-5.** Sound level meter. (Courtesy of H. H. Scott, Inc.)

**Figure 10-6.** Noise dosimeter. (Courtesy of Bausch & Lomb.)

exposure to noise as a function of noise level and time. This could also be done with a sound level meter, a stop watch, and a notebook, but it's much simpler and cheaper to use a dosimeter.

The dosimeter is a pocket-size monitor with a microphone that attaches to the collar or ear. At the end of the day, the day's noise exposure can be read out by plugging the dosimeter into a separate desk-top console readout unit.

For any serious noise measuring, it's necessary to analyze the jumble of sounds that make up a typical noise, as most noises are not pure tones. Octave band analyzers provide the frequency distribution of the noise. With this information it's possible to pinpoint noise sources which may not be obvious at all with a simple sound level meter.

Sound (noise) measuring equipment should be chosen carefully to avoid expensive duplication of functions and the need for highly trained operators. American National Standard ANSI S1.4-1971, the specification for sound level meters, specifies four types of sound level meters: Type 1, Precision; Type 2, General Purpose; Type 3, Survey; and Type S, Special Purpose.

Special impact noise analyzers are required to evaluate sudden or impact noises.

## NOISE CONTROL

If you're willing to pay the price, the means exist for controlling noise.

Acoustical tile is a much used standard treatment for certain noise problems but should not be considered the only or the ultimate solution. Other solutions may be better and cheaper (refer to Figures 10-7 and 10-8 and Tables 10-7, 10-8, and 10-9). The most obvious is to limit your time in noisy locations. For any extensive work to control noise, hire an acoustic engineer.

While it is true that noise is sound without value, there should be no attempt in noise control to eliminate all sound. A sound-proof room is an uncomfortable place; and even a very quiet room can be annoying if it is so quiet that the ordinary sounds of walking, talking, and breathing are distracting. Where too much soundproofing has been added to offices, acoustic engineers have had to generate a small amount of controlled, background noise to mask ordinary sounds which suddenly became objectionable.

In any noise control program, the first step is to take a sound level survey with the survey meter set on the *A*-scale at slow response. The survey will show precisely how bad the problem is and will often pinpoint the greatest sources of noise, which may not be the ones you suspected.

In a home survey, you may have to get by with inexpensive sound level meters, but in industry surveys, more sophisticated instruments, such as octave band analyzers, should be used. While expensive, they can pay for themselves by avoiding unnecessary or ineffective noise control techniques.

After the noise control materials and devices have been put into use, take another sound level survey to determine if the control measures actually worked.

Although noise in the home may not be hazardous to your health, it should be controlled or reduced because of its annoyance and disturbance.

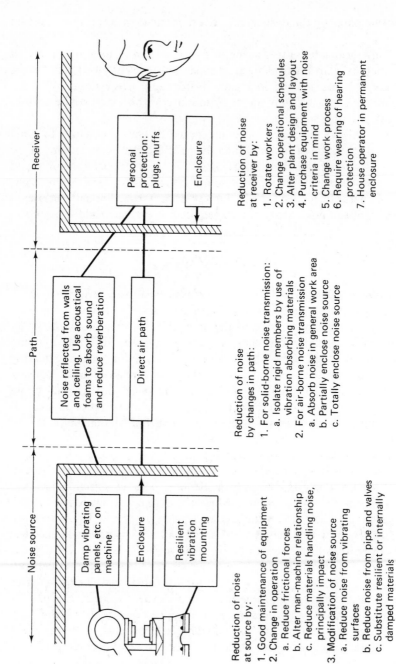

**Figure 10-7.** Noise system concept: source-path-receiver. (Courtesy of Tull Environmental Systems.)

Reduction of noise
at source by:
1. Good maintenance of equipment
2. Change in operation
  a. Reduce frictional forces
  b. Alter man-machine relationship
  c. Reduce materials handling noise,
     principally impact
3. Modification of noise source
  a. Reduce noise from vibrating
     surfaces
  b. Reduce noise from pipe and valves
  c. Substitute resilient or internally
     damped materials

Reduction of noise
by changes in path:
1. For solid-borne noise transmission:
  a. Isolate rigid members by use of
     vibration absorbing materials
2. For air-borne noise transmission
  a. Absorb noise in general work area
  b. Partially enclose noise source
  c. Totally enclose noise source

Reduction of noise
at receiver by:
1. Rotate workers
2. Change operational schedules
3. Alter plant design and layout
4. Purchase equipment with noise
   criteria in mind
5. Change work process
6. Require wearing of hearing
   protection
7. House operator in permanent
   enclosure

210

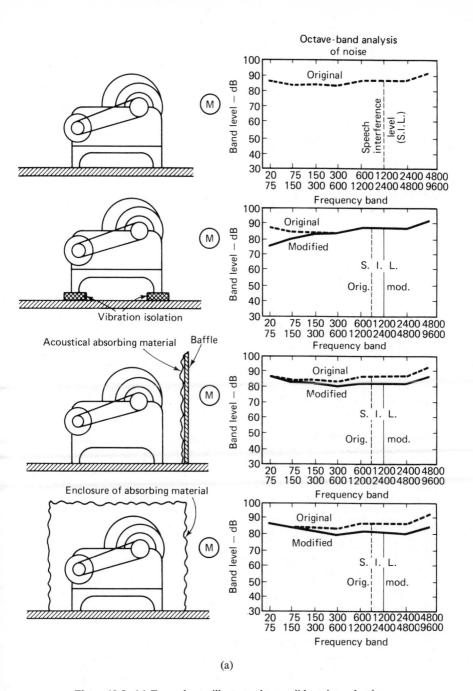

(a)

**Figure 10-8.** (a) Examples to illustrate the possible noise reduction effects of some noise control measures. (b) Examples to illustrate the noise reduction possible by the use of enclosures. (Courtesy of General Radio Co. From *Handbook of Noise Measurement*, 7th ed., 1972.)

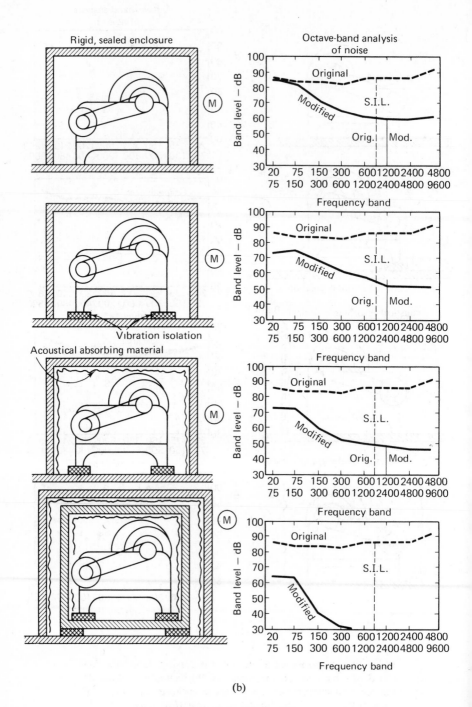

(b)

Figure 10-8. (Continued)

**TABLE 10-7** *Reducing noise at the source*[1]

| | Control Method | Examples | Solutions & Comments |
|---|---|---|---|
| **Maintenance** | Preserve original noise level through parts replacement, repair, conditioning, or adjustment. | Replace worn bearings or parts. Adjust loose fittings. Secure loose covers or shields. | Frequently cause of excessive noise levels on existing equipment. |
| **Operation** | Reduce frictional forces. | Lubricate. Use cutting oils and coolants. Use sharp cutting tools. | |
| | Alter man-machine relationship. | Change location or orientation of noise source when directionality of noise source is apparent. | Effective only for near field. Very little reduction in reverberant field. |
| | Reduce materials handling noise, principally impact. | Apply damping materials to chutes, hoppers, bins, conveyors, tote boxes, stock, guides, etc. | Use CON-PLEX, SOUNDEFENDER DP damping sheet; SHEALD® lead sheet. |
| **Modification** | Reduce noise from vibrating surfaces, such as machine enclosures, housings, guards, etc. | Apply damping material to surface for sound energy dissipation. | Use SOUNDEFENDER DP damping sheet or SOUNDEFENDER LF when damping plus absorption are needed. |
| | Reduce noise from pipe and valves. | Treat pipe noise when treating noise from compressors, pumps, etc. Often piping noise exceeds machine noise. Requires sound-absorbing material *plus sound barrier*. | Wrap with $1\frac{1}{2}''$ or $2''$ of fiberglass or mineral wool, then cover with 1 lb. per square foot SHEALD® lead, which provides sound barrier. Reductions of up to 55 dB in higher frequencies and 18 dB in lower frequencies are common. |
| | Substitute resilient or internally damped materials. | Use belt drives instead of gears. Nylon gears. | May not be practical due to heavy loads. |

®SHEALD is a Registered Trademark of Cominco, Ltd.
[1]Courtesy of Tull Environmental Systems.
 Note: Mention of specific brands and models in this table and in Tables 10-8 and 10-9 does not necessarily imply endorsement by the author or publisher.

**TABLE 10-8**  *Reducing Noise by changes in noise path*[1]

| | Control Method | Examples | Solutions & Comments |
|---|---|---|---|
| **Solid Borne Trans- mission** | Isolate rigid members by use of vibration absorbing materials. | Use flexible mounts or pads under machines or between parts where practical. Inertia blocks where heavy impact or shock is transmitted through foundation. | Install Unisorb vibra- tion isolating pads, Level-Rite machinery mounts. |
| **Air Borne Trans- mission** | Absorb noise in general work area. | Use absorbent materials on walls and ceilings. | *"Acoustisorber" Space Units* may be suspended from ceiling or fastened to walls. Acoustical foam sheets or blocks can be fastened to walls, partitions, etc. Maxi- mum overall noise reduction about 10 dB. |
| | Partially enclose noise source. | Use barriers between source and recipient of noise, can be of almost any size or shape. Walls, two and three sided enclosures, tun- nels, and ducts made from sound attenuating materials combined with sound absorbing materials. | *Tull Metal Panel Sys- tem* provides maximum sound reduction. Easily assembled and dis- assembled. *Tull Flexible Curtain Enclosures*, made from TLB-G5 leased vinyl and lined with foam are inexpensive, versatile, and effective. SOUNDEFENDER *LF* Lead/Foam applied to inside of metal or ply- wood enclosures are economical and effec- tive. Partial enclosures, depending upon indi- vidual conditions, can reduce noise up to 10 to 15 dB. |

[1]Courtesy of Tull Environmental Systems.

TABLE 10-8 *(Continued)*

| Control Method | | Examples | Solutions & Comments |
|---|---|---|---|
| **Air Borne Trans- mission** | Totally enclose noise source. | Completely contain noise source, using acoustical metal enclosure, with acoustical doors and windows. Baffle access openings. | *Tull Metal Panel* enclosures utilizing solid steel with perforated steel, and with acoustical absorbent material between, give high transmission loss plus absorbent capacity. Joining system, doors, and windows designed to give maximum integrity to structure. SOUNDEFENDER *LF* Lead/Foam applied to inside of metal or plywood enclosures are economical and effective. Noise reduction of 20 dB to 40 dB possible with proper design. Consideration must be given to ventilation and work flow. |

Unfortunately, modern home construction has emphasized light-weight materials, open spaces, poor workmanship, and high-pressure heating, cooling, and plumbing systems, all of which make noise problems worse. The following steps, which are based on Environmental Protection Agency suggestions, will provide a quieter home or workshop:

1. Close out the outside noises by closing the windows. Use air conditioning if necessary. Use caulking to plug the leaks.
2. Use noise-absorbing materials on floors, especially in high-traffic areas.
3. Use solid interior doors—not hollow-core. Weatherstrip interior doors as well as exterior ones. Use adjustable threshold gasket.
4. Hang heavy drapes over the windows closest to outside noise sources.
5. Put rubber or plastic treads on uncarpeted stairs.
6. Use upholstered rather than hard-surfaced furniture to deaden noise.
7. Install sound-absorbing ceiling tile in the kitchen. Wooden cabinets will vibrate less than metal ones.

**TABLE 10-9**   *Reducing noise at the receiver* [1]

| | Control Method | Examples | Solutions & Comments |
|---|---|---|---|
| **Personnel** | Rotate workers. | Put limitations on length of exposure, based on damage risk criteria. | Possible conflict with skilled labor availability and union work rules. |
| | Change operational schedules. | Schedule noisy operations for minimum time exposure. | Not usually practical in process oriented plants. |
| **Plant and Equipment** | Alter plant design and layout. | Segregate noisy processes in one area. | Frequently existing structural elements can be used as barriers. |
| | Purchase equipment with noise criteria in mind. | Include noise criteria in specifications. | Best long term approach to noise control. Costs less than subsequent modification. |
| | Change work process. | Weld instead of riveting. Press instead of forging. | A high cost approach. Not always feasible. |
| **Personal** | Require wearing of hearing protection devices. | Ear plugs, ear muffs. | Tull E-A-R® plugs provide protection, comfort, and universal fit. Do not require individual fitting. |
| **Structural** | House operator in permanent enclosure. | Operator monitors instruments or equipment from within quiet room. | *Tull Metal Quiet Rooms* are an effective means by which noise reaching workers is greatly reduced. |

[1]Courtesy of Tull Environmental Systems.
®E-A-R is a Registered Trademark of National Research Corp.

8. Line heating/air conditioning ducts with fiberglass.

9. Wrap water pipes with soft fiberglass.

10. Use a foam pad under blenders and mixers.

11. Use insulation and vibration mounts when installing dishwashers or any other vibrating equipment.

12. Install washing machines in the same room with heating and cooling equipment, preferably in an enclosed space.

13. Add extra wall boards to walls to increase rigidity.

14. Use earphones to listen to stereo or TV.

15. Compare the noise outputs of different makes of an appliance before making your selection.

Modification costs for noise insulating a house are given in Table 10-10.

**TABLE 10-10**  *Bolt Beranek and Newman's estimate of the probable range of modification costs for a 1,000 square foot house, 1966 (Exclusive of costs for ver.tilation)*[1]

| House Type | Noise Insulation Improvement 5-10 PNdB | 10-15 PNdB | 15-20 PNdB |
|---|---|---|---|
| Light Exterior Walls | $260 | $1,600 | $4,000 |
| (wood, metal, stucco, | to | to | to |
| or composition) | $820 | $2,400 | $4,500 |
| Heavy Exterior Walls | $260 | $1,600 | $2,800 |
| (brick, masonry, or | to | to | to |
| concrete block) | $820 | $2,400 | $3,400 |

[1]Environmental Protection Agency, *The Economic Impact of Noise*, NTID300.14.

## EAR (HEARING) PROTECTION

When noise control techniques are inadequate and sound levels cannot be reduced to the OSHA maximum limits, then personal protective equipment (earplugs or earmuffs or both) must be provided and used by all exposed workers to lower the noise to the inner ear to acceptable levels. (Another approach is to simply limit the amount of time you spend around excessive-noise generators.) Ear protectors should be used only as the *last* line of defense and even then should be considered an interim measure till noise suppression techniques can be improved.

Some people find hearing protectors inconvenient, heavy, cumbersome, hot, and possibly irritating to the ear canal. Because of these problems, it's imperative that protectors be chosen carefully, in order to minimize the discomfort and to ensure adequate protection. Even then, some employees (including yourself?) must be convinced that the protectors are for their own good.

Despite the fear of some who believe that earplugs or muffs can mask important sounds, it's been found that a worker wearing protectors in an area of high noise is better able to hear warnings or instructions than without them.

Ordinary cotton stuffed in the ears offers so little attenuation that it is considered unacceptable. Wax-impregnated cotton, on the other hand, may be used if it is provided fresh daily and inserted properly.

Plugs are inexpensive and effective but generally must be fitted to each user. And only a trained person under the direction of a physician should issue most plugs. Com-Fit Hearing Protector (Figure 10-9), a product of

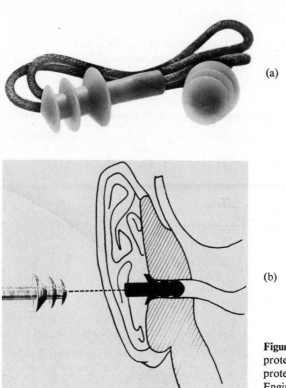

(a)

(b)

**Figure 10-9.** (a) Com-Fit hearing protector inserted. (b) Com-Fit protector. (Courtesy of Sigma Engineering, Norton Company.)

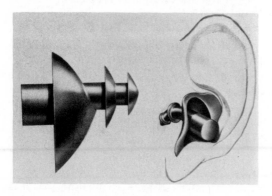

**Figure 10-10.** Auri-Seal hearing protector. (Courtesy of Sigma Engineering, Norton Company.)

Sigma Engineering Company, is a silicone ear insert that uses a triple flange design to block both high and low frequency noise. The Auri-Seal (Figure 10-10) hearing protector is a variation of this insert, which adds a flange to fill the outer part of the ear. The Sonic Ear-Valv (Figure 10-11) is a tiny ear insert device, which, the manufacturer says, reduces high frequency impulse-impact noise and permits the passage of normal, low frequency

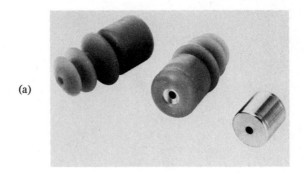

(a)

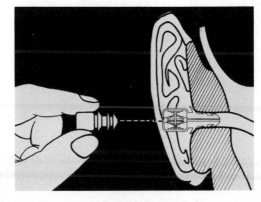

(b)

**Figure 10-11.** (a) Sonic Ear-Valv. (b) Ear-Valv inserted. (Courtesy of Sigma Engineering, Norton Company.)

sounds such as conversation. Unlike conventional ear plugs, the Ear-Valv does not seal the ear canal.

Some protectors, such as Sound Sentry (Figure 10-12), can be applied by the wearer, according to the advertisements, without the need for personal fitting by a doctor, nurse, or hearing technician.

Earmuffs (Figure 10-13) cost more but require no special fitting. Long hair or eye glasses may interfere with the seal and reduce the attenuation specified by the manufacturer.

Typical noise level reduction figures as determined by the U.S. Air Force for hearing protection equipment are given in Table 10-11. The Department of Labor advises that manufacturers' claims for specific attenuation values are determined under ideal conditions; it's best to assume that the attenuation actually attained on the job will be at least 5 dB less than the stated value. Whatever device you use, make sure it has been tested in accordance with American National Standard Z 24.22-1957(R-1971). Air Force regulation AFR 160-3 recommends ear protection for the individual who is continuously exposed to sound pressure levels above 85 dB. OSHA requires

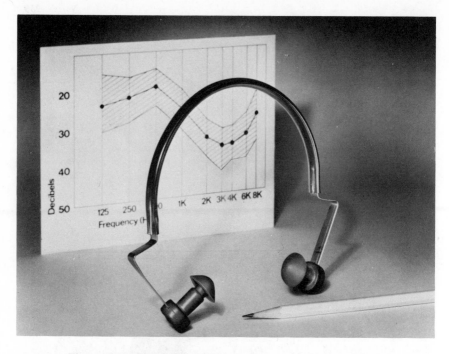

**Figure 10-12.** Sound Sentry hearing protector. (Courtesy of H. E. Douglass Engineering Sales Co.)

**Figure 10-13.** Ear muff hearing protector. (Courtesy of Bausch & Lomb.)

**TABLE 10-11**   *Average octave band noise level reduction in ear canal achieved by use of personal protective equipment* [1]

| Types of Equipment Used | BANDS | | | |
|---|---|---|---|---|
| | 300-600 cps | 600-1200 cps | 1200-2400 cps | 2400-4800 cps |
| Headset Earphone Covers | −7 dB | −13 dB | −20 dB | −30 dB |
| Standard Ear Plug or Ear Muff used alone | −14 dB | −18 dB | −25 dB | −30 dB |
| Standard Ear Plug and Ear Muff used together | −24 dB | −28 dB | −37 dB | −40 dB |

[1]Air Force regulation AFR 160-3

protection above 90 dB. By AF regulations, at levels of 130 to 140 dB, a man must wear earmuffs *and* ear plugs.

## ULTRASONICS

Ultrasonic emissions are generally considered noise, even though their high frequency (20,000 Hz or more) makes them inaudible. As shown in Table 10-12, ultrasonic medical diagnostic and therapeutic units are widely

**TABLE 10-12**   *Estimates of inventory of ultrasonic devices, Jan. 1970*[1]

| Device | Estimated Number |
|---|---|
| Medical diagnostic units | 3,000 |
| Medical therapeutic units | 3,000 |
| Cleaners | Tens of thousands |
| Welders | Few thousand |

[1]*Electronic Product Radiation and the Health Physicist* Bureau of Radiological Health Publication BRH/DEP 70-26.
(Courtesy of Bureau of Radiological Health, Food and Drug Administration, Dept. of Health, Education, and Welfare.)

used, to say nothing of ultrasonic cleaners. At this time, ultrasonic equipment does not appear to be hazardous *when used as directed*. However, the lack of knowledge of the biological hazards of ultrasonics is a matter of great concern to scientists. At intense levels, ultrasonics can disintegrate human cells and enzymes, but at lower levels the risk is not known. Some workers using ultrasonic cleaners and drills have complained of headaches and nausea. The Russians recommend that persons with hearing problems or nervous disorders should not be employed around ultrasonic equipment. Until more specific safeguards can be determined, avoid contact with ultrasonic equipment during operation and keep high level ultrasonic equipment shielded during operation (see Table 10-13).

**TABLE 10-13**[2]  *Acceptable sound pressure levels for ultrasonic cleaners*[1]

### FREQUENCY RANGE 175 to 5,600 Hz

| Octave Band Center Frequency | Octave Band Sound Pressure Level in dB |
|---|---|
| 250 Hz | 94 dB |
| 500 Hz | 88 dB |
| 1,000 Hz | 85 dB |
| 2,000 Hz | 84 dB |
| 4,000 Hz | 84 dB |

### FREQUENCY RANGE 5,600 to 18,000 Hz

| Center Frequency of the One-third Octave Band | One-Third Octave Band Sound Pressure Level in dB |
|---|---|
| 6,300 Hz | 80 dB |
| 8,000 Hz | 80 dB |
| 10,000 Hz | 80 dB |
| 12,500 Hz | 80 dB |
| 16,000 Hz | 80 dB |

### FREQUENCY RANGE 18,000 to 45,000 Hz

| Center Frequency of the One-third Octave Band | One-Third Octave Band Sound Pressure Level in dB |
|---|---|
| 20,000 Hz | 105 dB |
| 25,000 Hz | 110 dB |
| 31,500 Hz | 115 dB |
| 40,000 Hz | 115 dB |

[1]All sound pressure levels are expressed in decibels (dB) reference 0.0002 dynes per square centimeter (0.0002 microbar).

These are the maximum acceptable octave band levels, when measured at the head position (ears) of the operator, when at the normal operator's position by the ultrasonic cleaner, and when the cleaner is the source of the sound. The sound pressure level of each octave band shall be at least 10 dB lower when measured at any position about the cleaner located a radius length of 15 feet from the position of the center of the operator's head.

[2]*Present Federal Control of Health Hazards from Electronic Product Radiation and Other Types of Ionizing Radiation* Bureau of Radiological Health, November 1970.
(Courtesy of Bureau of Radiological Health, Food and Drug Administration, Dept. of Health, Education, and Welfare.)

## SONAR

All operating sonar equipment is capable of causing physiological injury to divers and swimmers in their vicinity, according to Navy manual NAVMAT P5100. To prevent such injuries, stay far away, unless you have access to certain classified Navy documents which give the specific safe distances for swimmers and divers.

# *11*

# TOOLS

As far as tools are concerned, an electronics workshop is not a particularly dangerous place. Generally there are no grossly unsafe tools such as power saws, routers, planers, etc., that you would find in a woodworking shop. In fact, about the only power tools to be found in most electronics shops are the ordinary drill and the grinding wheel. With reasonable precautions, neither one is particularly unsafe. But if you ignore basic safety, you can easily get hurt. For example, if safety goggles aren't used with the grinding wheel, it's possible to be blinded by flying chips either from the wheel itself or from the tool you are sharpening.

Fortunately, most electronic shop injuries are relatively minor: burns from soldering irons, skinned knuckles from slipping wrenches or screwdrivers, gouged or mashed fingers.[1] To avoid such injuries, you should take the following precautions:

1. Keep your work area off limits to unauthorized guests (and that's anyone, especially children, you can't trust with tools). If for some reason you can't keep your work area private, arrange to lock up your tools or store them out of reach (Figure 11-1). If you have stationary power tools, connect them to key switches and keep the key in your pocket.

2. Arrange and maintain your shop for comfort. A comfortable shop

---

[1]Hand tool injuries may be minor, but *Accident Facts* reports that the average cost of a compensable hand tool injury is $850.

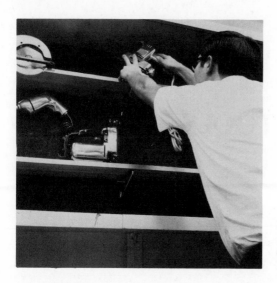

**Figure 11-1.** Keep tools in a safe, dry place and out of the reach of children. (Courtesy of Power Tool Institute.)

will always be safer than one that is either too crowded, poorly lighted, inadequately ventilated, too hot or too cold, or that has dirty, slippery floors. A sturdy workbench wide enough for equipment and tools and at the proper height for your comfort is essential. If you can arrange it, keep both ends of the workbench floor area clear so that you can walk around the ends for easier access to awkward-to-reach equipment. Whether you sit or stand, you should be comfortable— either cushion your chair or use a nonconducting floor mat to stand on. Because of the shock hazards, avoid metal chairs and work-benches and also damp shops.

3. Avoid clutter (Figure 11-2). A cluttered workbench is neither efficient nor safe. Arrange convenient, easy-to-reach dry storage racks, hold-ers, drawers, and cabinets for all of your tools. Mark the tool loca-tions so that you can return your tools easily to the right place without searching. Proper storage will keep your tools from being abused and will keep cutting tools sharp. Obviously, the storage technique should not be a safety hazard in itself; for example, don't hang screwdrivers where you may rake your hand across their sharp points as you reach for another tool. If you keep tools in a tool box, store sharp-edged tools so they won't nick you as you're rummaging through the tool box.

4. Dress properly. Long-sleeved shirts, neckties (except for bow ties), badge necklaces, work gloves, wrist watches, bracelets, rings, and long hair catch on tools easily, resulting in accidents (Figure 11-3). Jewelry creates an additional shock hazard near electrical circuits;

**Figure 11-2.** A cluttered workbench is an invitation to trouble. (Courtesy of Power Tool Institute.)

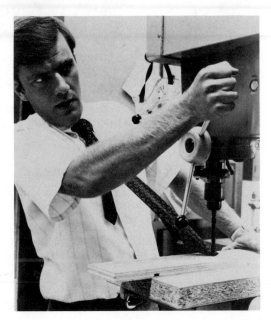

**Figure 11-3.** Loose clothing or jewelry can get caught in moving parts of power tools. (Courtesy of Power Tool Institute.)

it can also be the cause of serious burns of the fingers and wrists as a result of shorting out a circuit. Even metal-rim glasses may cause electrical shocks.

5. Wear safety goggles or a face shield (Figure 11-4) on any job that involves grinding, chipping, drilling, or the use of compressed air. Incidentally, when using compressed air to remove dust or dirt from a chassis, do not inadvertently aim the air hose at another person and do not use the air to remove dust from your clothes.

**Figure 11-4.** Safety goggles and machine guard protect this man. Bracelet, however, is a hazard. (Courtesy of Power Tool Institute.)

6. Keep your tools properly sharpened, clean, lubricated, and adjusted. (But don't try cleaning or oiling power tools while they're running.) A rusty, dull tool may cause you to use undue force, slipping or catching at the wrong place.

7. Inspect tools before you use them—broken, cracked, or abused tools can leave *you* broken, cracked, and abused. When in doubt, discard the tool. For the cost of a new screwdriver, for example, it's just not worth the risk of using one with a cracked handle when you're making an adjustment on a high voltage circuit.

8. Use only the tools intended for a particular job; a screwdriver was not designed to be used as a pry bar or a chisel. Pliers should not be used where wrenches are called for. Wrenches should not be used as substitute hammers. Ordinary screwdrivers should not be used on Phillips screws. Using the proper tool makes your work easier and faster as well as safer. There are an amazing number of specialized tools available to the technician today: tools for installing and replacing flat-pack integrated circuits, miniature tools, crimping tools, special vises, and chassis cradles. Consult your tool catalog.

9. Don't force a tool to do a job that is too big for it (Figure 11-5). That is, don't try to exceed a tool's rating by adding pipes for leverage. To do so is to ask for trouble: the tool is likely to break and go flying.

**Figure 11-5.** Forcing a tool to do more than it was designed for can cause workshop accidents. (Courtesy of Power Tool Institute.)

10. No matter what tool you are using, hold it so that if it slips it will not damage you, the equipment, or anyone else.

11. Don't leave tools on ladders, throw them to people overhead, or carry them in your pockets or hands while climbing ladders. Pull tools up with a rope, but don't let anyone stand underneath during the process. Or carry them in specially designed leather tool pockets with the sharp points of the tools pointing down (Figure 11-6).

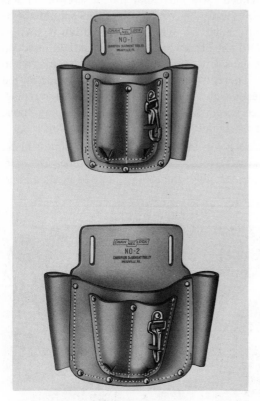

**Figure 11-6.** Tool holster allows tools to be carried safely. (Courtesy of Channellock Co.)

12. In flammable atmospheres, use nonsparking tools made from nonferrous alloys to prevent explosions from impact sparks.
13. Use Clever Kleps test probes (from Rye Industries) to measure high voltage without getting your hands near the high voltage.
14. In equipment construction, try to avoid sharp corners and burrs on equipment chassis; they're obvious hand slicers.
15. Keep a fire extinguisher handy.

## SPECIAL TOOL PROBLEMS

*Chisels and punches.* If they have mushroomed heads, replace them. Don't hit a cold chisel with an ordinary claw hammer; use a ball peen hammer instead. Wear safety goggles when using a cold chisel.

*Drills.* Clamp material to be drilled, if possible. Remove chuck key before operating drill.

*Files.* Always keep a protective handle on the sharp end (tang).

*Grinding wheel.* Use safety goggles or face shield. Try to use a grinder that has safety housing covers for the grinding wheel; also use eye shields. Before starting, make sure the wheel is not cracked or damaged. Stand to the side until the grinder comes up to speed. Do not apply work too quickly to a cold wheel and use light pressure when you start grinding.

*Knives.* Keep blade(s) sharp (Figures 11-7 and 11-8). Store and carry in protective sheath. Cut away from you. Consider using wire strippers instead.

**Figure 11-7.** Skinning knife—harmless when used properly; dangerous when abused. (Courtesy of Channellock Co.)

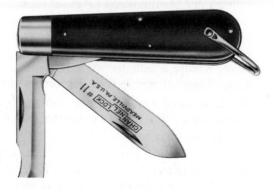

**Figure 11-8.** Electricians' knife—important to keep it dry and sharp. (Courtesy of Channellock Co.)

*Metal-cutting snips.* Keep snips oiled and adjusted. Don't attempt to cut material that's obviously too heavy for your particular snips.

*Microflame ® Gas Welding Torch* (Figure 11-9). A useful gadget for welding, brazing, and soldering, but beware of its 5,000° F flame. Make sure no flammable materials or aerosol cans are nearby while you're

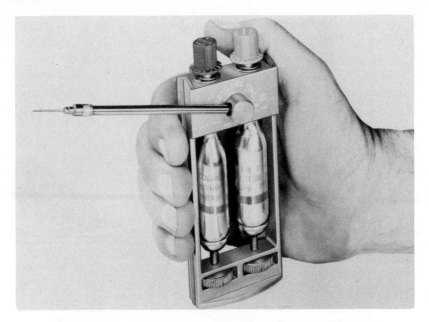

**Figure 11-9.** Microflame ® gas welding torch. (Courtesy of Micro-flame Corp.)

using such torches. Also, make sure an approved fire extinguisher is handy. The Microflame Company gives the following warnings with the torch: When cutting, brazing, or welding, it is advisable that you wear dark glasses. Keep Micronox and LP gas away from open flame. Store cylinders of Micronox and LP gas at room temperature (not over 130°F). Keep away from direct sunlight, radiators, stoves, hot water and other heat. Do not incinerate cylinders. LP gas is flammable. Do not store LP gas in a room used for habitation. Keep cylinders away from children. Never remove a cylinder while torch is in operation or near an open flame. Never attempt to light the torch when one cylinder is removed.

*Powder-actuated fastening tools.* Only properly trained and qualified operators should be allowed to operate such tools. People have been killed through improper use of these tools.

*Screwdrivers.* Make sure the blade fits the screw, neither too long or too short, neither too thin or too wide. Regrind the tips when they become worn, bent, or chipped. Use only properly insulated screwdrivers for electrical work (Figure 11-10).

*Side-cutting pliers.* When cutting wire, turn pliers so that any flying pieces of wire will not hit anyone. Cut wire at 90 degree angle. If there's any possibility the wire will head in your direction, put on safety goggles. Do

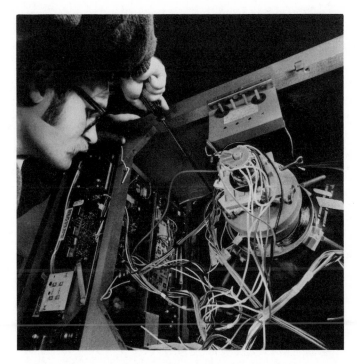

**Figure 11-10.** In a situation like this, a well-insulated screwdriver can prevent bad shock. (Courtesy of Xcelite Corp.)

not depend on ordinary plastic insulation on pliers to prevent shock; get properly insulated pliers for cutting live wires.

*Soldering iron.* Use a stand (Figure 11-11) or cradle to prevent accidental burns. Keep aerosols and flammable materials away. Be extremely wary of using any solder that contains cadmium.

*Staple-gun tackers.* Useful for telephone, communications, and electronics wire and cable installations. If they are aimed carelessly, they can fire staples with enough force to give serious eye injuries and other painful puncture wounds.

*Star drills.* Essential for drilling holes that are too large for your electric drill in concrete blocks, bricks, etc. In using, strike the drill with a heavy ball peen or sledge hammer (not a carpenter's hammer) and rotate the drill after each blow. Wear goggles to protect your eyes from flying chips of masonry. Make sure you will not be cutting into electric lines; if there's any doubt, pull circuit breakers or fuses, or wear insulated (rubber) gloves protected by regular gloves to prevent electric shock. To protect your hands, hold the drill with pliers.

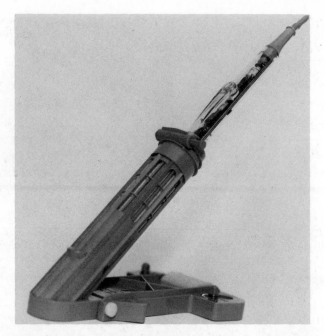

Figure 11-11. A soldering iron stand or cradle can prevent accidental burns. (Courtesy of the Edsyn Co.)

*Welding.* Safety precautions concerning welding are beyond the scope of this book. However, guard against infrared and ultraviolet radiation, burns from contact with hot metal or sparks, metal fumes and gases, and electric shock.

*Wrenches.* Much better than pliers for loosening or tightening nuts. When you must use appreciable force, pull on the wrench instead of pushing. If you must push, keep your hand open.

## POWER TOOLS

The following safety rules for power tools have been summarized from precautions developed by the Power Tool Institute. They should be used together with the rules previously stated (parenthetical remarks added by the author).

1. Know the tool you are using—its application, limitations, and potential hazards. Read instruction manuals (Figure 11-12).
2. If you have a choice, use double-insulated tools in preference to grounded tools. If a tool is equipped with a three-prong plug, it

**Figure 11-12.** Not many people make this mistake of throwing away the instruction manual. But failure to read the manual is just as much a mistake. (Courtesy of Power Tool Institute.)

should be plugged into a three-hole electrical receptacle. If an adapter is used to accommodate a two-prong receptacle, the adapter wire must be attached to a *known ground.* (Because of electric shock hazard, avoid standing on metallic or wet surfaces *even if* you're using well-grounded tools or double-insulated tools.)

3. Remove adjusting keys and wrenches before turning on the tool.

4. Keep guards in place and in working order. Do not remove or wedge out of the way.

5. Always be alert to potential hazards in your working environment, such as damp locations, the presence of highly combustible materials (gasoline, naptha), etc.

6. Avoid accidental startup. Make sure switch is off before plugging in cord, or when power is interrupted. Don't carry plugged-in tool with your finger on the switch.

7. Use only recommended accessories. Follow manufacturer's instructions.

8. Do not force tool. It will do a better and safer job at its designed speed.

9. Do not overreach. Keep proper footing and balance at all times.

10. Never leave a running tool unattended. Don't leave until it comes to a complete stop and is disconnected from its power source.

11. Don't surprise or touch anyone operating a power tool. The distraction could cause a serious accident.

12. Never adjust, change bits, blades, or cutters with the tool connected.

13. Secure your work. Use clamps or a vise to hold work, if possible. It frees both hands to operate the tool.

14. Do not use a tool with a frayed cord. (Better to replace the cord than to splice it. For three-wire tools make absolutely certain that wires are correctly replaced; faulty installation of the cord may cause the frame of the tool to be electrically hazardous.) Use only heavy-duty UL-listed extension cords of proper wire size and length. (If you have three-wire tools, get rid of all two-wire extension cords in your shop so you won't be tempted to use the wrong cord.)

15. Never brush away chips or sawdust while tool is operating.

16. Do not attempt field repairs. (If you must, be sure you know what you're doing. When repairing double insulated tools, use exact replacement parts.)

17. Inspect tools periodically for safety.

# *12*

# PRODUCT SAFETY

In any given year, an estimated 20 million Americans are injured as a result of accidents connected with household consumer products, according to the National Commission on Product Safety. Of this total, and excluding deaths and injuries from automobile accidents, a frightening 110,000 are hospitalized and 30,000 are killed.

Although the consumer product involved in each of these incidents is not always the prime cause of the mishap, it is too frequently the culprit. For evidence of this, observe the massive recall campaigns conducted by American manufacturers in recent years to repair hazardous products before they caused further death and injury. Automobile recalls have occurred so often that they no longer attract attention. Recall of electronic products, while not as common, is on the increase; in 1973, for example, one TV manufacturer recalled 52,000 color sets because of a potential shock hazard caused by a missing transistor.

If you manufacture, sell, or repair any consumer electronic products you should be aware of your legal obligations, which are given in this chapter.

In recent years consumer product hazards of fire, shock, and X-ray and microwave radiation have received extensive publicity in newspapers and TV reports throughout the nation. As the public becomes more and more aware of product safety, more and more people will insist on absolutely safe products. If they receive products which are unsafe, they may sue all those involved—the manufacturer, the distributor, and the retail store. So, let the seller beware!

The number of consumer products has been increasing dramatically, thereby increasing the consumer's exposure to risk of injury. Furthermore, previously acceptable risk levels may no longer be reasonable in light of available safety technology.

Just how much risk should the consumer have to assume? How much safety is reasonable? Is it possible to define "unreasonable hazard"? Professor Corwin D. Edwards has given this definition (From *Safety in the Marketplace—A Program for the Improvement of Consumer Product Safety*, April 1973):

> Risks of bodily harm to users are not unreasonable when consumers understand that risks exist, can appraise their probability and severity, know how to cope with them, and voluntarily accept them to get benefits that could not be obtained in less risky ways. When there is a risk of this character, consumers have reasonable opportunity to protect themselves; and public authorities should hesitate to substitute their value judgments about the desirability of the risk for those of the consumers who choose to incur it.
>
> But preventable risk is not reasonable (a) when consumers do not know that it exists; or (b) when, though aware of it, consumers are unable to estimate its frequency and severity; or (c) when consumers do not know how to cope with it, and hence are likely to incur harm unnecessarily; or (d) when risk is unnecessary in . . . that it could be reduced or eliminated at a cost in money or in the performance of the product that consumers would willingly incur if they knew the facts and were given the choice.

According to a Senate committee of the 92nd Congress (Consumer Safety Act of 1972, Report No. 92–749, April 13, 1972):

> A determination of unreasonable risk must be predicated upon the following factors: 1) the degree of the anticipated injury; 2) the frequency of such injury; 3) the effect upon the performance or availability of a product when the degree of the anticipated injury or the frequency of such injury is reduced; and 4) an evaluation of the utility of the product, in absolute terms and in varying modes of risk.

Improved product safety may be obtained by changing (1) the products, (2) the ways in which consumers use them, and (3) the environment in which they are used.

Of these solutions, the second is probably the least productive. "The prospects for measureable reform of human behavior," states the National Commission on Product Safety, "are distant." Or, as Senator Moss (Utah) puts it, "It is easier to modify things than people."

However, the Senate committee observed, neither self-interest nor competition has impelled manufacturers to produce products that are safe. Competition and voluntary actions of businessmen do not always suffice to safeguard the public interest. Competition does not inevitably take the form of a rivalry to produce the safest product. Indeed the competitive struggle may sometimes lead to a shaving of the costs of manufacture, involving some sacrifice of safety.

Because of these factors, Congress has had to pass several laws, as shown in Figure 12-1 and Table 12-1, to force manufacturers to build safer products. Probably the most important of the consumer protection laws is the Consumer Product Safety Act, which is described in the following paragraphs.

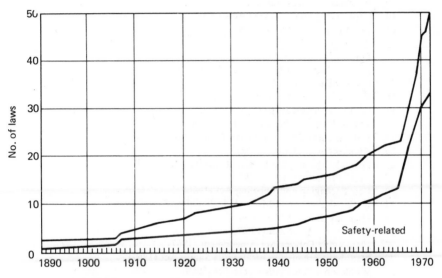

**Figure 12-1.** Federal consumer protection laws, 1890–1972. (From the Subcouncil on Product Safety of the National Business Council for Consumer Affairs, *Safety in the Marketplace—a Program for the Improvement of Consumer Product Safety*, Apr. 1973.)

### CONSUMER PRODUCT SAFETY ACT— PUBLIC LAW 92-573

The Consumer Product Safety Act, which went into effect in 1972, established an independent regulatory commission known as the Consumer Product Safety Commission to administer product safety laws. At the same time, the act gave the commission powerful enforcement tools: fines up to $500,000 and jail sentences for violators.

In passage of this act, Congress noted that:

1. An unacceptable number of consumer products that present unreasonable risks of injury are distributed in commerce.
2. Complexities of consumer products and the diverse nature and abilities

**TABLE 12-1** *Federal consumer protection laws, 1890–1972*[1]

1. Sherman Anti-Trust Act (1890)
2. Meat Inspection Acts (1890, 1906, 1907)
3. Clayton Act (1914)
4. Federal Trade Commission Act (1914)
5. Federal Power Act (1920)
6. Packers and Stockyard Act (1921)
7. Federal Communications Commission of 1934
8. Securities Exchange Act of 1934
9. Robinson-Patman Act (1936)
10. Food, Drug and Cosmetic Acts (1938)
11. Wool Products Labeling Act of 1939
12. Public Health Service Act (1944)
13. Federal Insecticide, Fungicide and Rodenticide Act (1947)
14. Fur Products Labeling Act (1951)
15. Flammable Fabrics Act (1953)
16. Refrigerator Safety Act (1956)
17. Poultry Products Inspection Act (1957)
18. Textile Fiber Products Identification Act (1958)
19. Federal Hazardous Substances Labeling Act (1960)
20. Kefauver Act (Drugs) (1962)
21. Federal Cigarette Labeling and Advertising Act (1965)
22. Child Protection Act of 1966
23. Fair Packaging and Labeling Act (1966)
24. National Traffic and Motor Vehicle Safety Act of 1966
25. Highway Safety Act of 1966
26. Wholesome Meat Act (1967)
27. Flammable Fabrics Act Amendment (1967)
28. National Commission on Product Safety Act (1967)
29. Wholesome Poultry Products Act (1968)
30. Consumer Credit Protection Act (1968) (Truth-in Lending)
31. Radiation Control for Health and Safety Act of 1968
32. Fire Research and Safety Act of 1968
33. Natural Gas Pipe Line Safety Act of 1968
34. Child Protection and Toy Safety Act of 1969
35. Public Health Cigarette Smoking Act of 1969
36. Securities Investor Protection Act of 1970
37. Comprehensive Drug Abuse Prevention and Control Act of 1970
38. Egg Products Inspection Act (1970)
39. Fair Credit Reporting Act of 1970
40. Postal Reorganization Act (1970) (covers unordered merchandise)
41. Poison Prevention Packaging Act of 1970
42. Lead-Based Paint Poison Prevention Act (1970)
43. Occupational Safety and Health Act of 1970
44. Federal Boat Safety Act (1971)
54. Drug Listing Act of 1972
46. Motor Vehicle Information and Cost Savings Act (1972)
47. Consumer Product Safety Act of 1972

[1]The Sub-Council on Product Safety of the National Business Council for Consumer Affairs, *Safety in the Marketplace—A Program for the Improvement of Consumer Product Safety*, April 1973.

of consumers using them frequently result in an inability of users to anticipate risks and to safeguard themselves adequately.
3. The public should be protected against unreasonable risks of injury associated with consumer products.

Pertinent parts of this act have been extracted in the following paragraphs:

*Purposes*
The purposes of this act are to:
1. Protect the public against unreasonable risks of injury associated with consumer products.
2. Assist consumers in evaluating the comparative safety of consumer products.
3. Develop uniform safety standards for consumer products and to minimize conflicting state and local regulations.
4. Promote research and investigation into the causes and prevention of product-related deaths, illnesses, and injuries.

*Definitions*
*Consumer product.* Any article or component part thereof, produced or distributed (1) for sale to a consumer for use in or around a permanent or temporary household or residence, a school, in recreation, or (2) for the personal use, consumption or enjoyment of a consumer in or around a permanent or temporary household or residence, a school, in recreation, or otherwise. (In the following discussion, "consumer products" will be considered to be the same as "distributed in commerce.")
*Risk of injury.* A risk of death, personal injury, or serious or frequent illness.
*Manufacturer.* Any person who manufactures or imports a consumer product.
*Distributor.* A person to whom a consumer product is delivered or sold for the purpose of distribution in commerce, except that such term does not include a manufacturer or retailer of such product.
*Retailer.* A person to whom a consumer product is delivered or sold for purpose of sale or distribution by such a person to a consumer.
*To distribute in commerce* and *distribution in commerce.* To sell in commerce, to introduce or deliver for introduction into commerce, or to hold for sale or distribution after introduction into commerce.

*Consumer Products Safety Standards.* The Commission may issue consumer product safety standards which may specify requirements for performance, composition, contents, design, construction, finish, or packaging of a consumer product. These standards may also require consumer products to be marked with or accompanied by clear and adequate warnings or instructions.

*Banned Hazardous Products.* Whenever the Commission finds that (1) a consumer product presents an unreasonable risk of injury and (2) no feasible consumer product safety standard under this act would adequately protect the public from the unreasonable risk of injury associated with such product, the Commission may ban such products as hazardous.

*Stockpiling.* The Commission may prohibit a manufacturer of a consumer product from stockpiling any product to which a consumer product safety rule applies, so as to prevent such manufacturer from circumventing the purpose of such consumer product safety rule.

*Amendments.* Any interested person or consumer organization may petition the Commission to issue, amend, or revoke a consumer product safety rule.

*Imminent Hazards.* The Commission may obtain court action against (1) an imminently hazardous consumer product for seizure of such product, (2) any manufacturer, distributor, or retailer of such product, or (3) both. The term *imminently hazardous consumer product* means a consumer product that presents imminent and unreasonable risk of death, serious illness, or severe personal injury.

*New Products.* The Commission may prescribe procedures to insure that the manufacturer of any new consumer product furnishes notice and a description of such product to the Commission before distributing it.

The term *new consumer product* means a consumer product that incorporates a design, material, or form of energy exchange which (1) has not previously been used substantially in consumer products and (2) as to which there exists insufficient information to determine the safety of such product.

*Product Certification and Labeling.* Every manufacturer of a product which is subject to this act must certify that his product conforms to all applicable consumer product safety standards, and must specify any applicable standards. Such certificate shall accompany the product or shall otherwise be furnished to any distributor or retailer to whom the product is delivered.

*Notification and Repair, Replacement, or Refund.* The term *substantial product hazard* means:

1. a failure to comply with an applicable consumer product safety rule that creates a substantial risk of injury to the public, or
2. a product defect that (because of the pattern of defect, the number of

defective products distributed, the severity of the risk, or otherwise) creates a substantial risk of injury to the public.

Whenever a manufacturer (or distributor or retailer) of a consumer product determines that such product:

1. fails to comply with an applicable consumer product safety rule; or
2. contains a defect that could create a substantial product hazard,

he must immediately inform the Commission of such failure to comply or of such defect, unless he knows that the Commission has been adequately informed of such defect or failure to comply.

If the Commission determines that a product presents a substantial product hazard and that notification is required in order to adequately protect the public from such substantial product hazard, the Commission may order the manufacturer or any distributor or retailer of the product to:

1. Give public notice of the defect or failure to comply.
2. Mail notices to each person who is a manufacturer, distributor, or retailer of such product.
3. Mail notices to those to whom it is known that such product was delivered or sold.

If the Commission determines that a product presents a substantial product hazard, it may order the manufacturer, distributor, or retailer of the product to: (1) repair the defect in the product, (2) replace it with one that does not contain the defect, or (3) refund the purchase price.

*Inspection and Recordkeeping.* Commission employees are authorized to enter and inspect any factory, warehouse, or establishment in which consumer products are manufactured or held, or any vehicle being used to transport consumer products.

Manufacturers, private labelers, and distributors of consumer products must establish and maintain such records, make such reports, and provide such information as the Commission may require.

*Imported Products.* Any imported consumer product shall be refused admission to this country if it:

1. Fails to comply with an applicable consumer product safety rule.
2. Is not accompanied by a certificate.
3. Is not labeled in accordance with regulations.
4. Is or has been determined to be an imminently hazardous consumer product.

5. Has a product defect which constitutes a substantial product hazard.

*Prohibited Acts.* It shall be unlawful for any person to:

1. Manufacture for sale, offer for sale, distribute in commerce, or import into the United States any consumer product that is not in conformity with an applicable consumer product safety standard under this act or that has been banned as hazardous.
2. Fail or refuse to permit access to or copying of records, or fail or refuse to make reports or provide information.

*Civil Penalties.* Any person who knowingly violates the Prohibited Acts section of this act shall be subject to a civil penalty not to exceed $2,000 for each such violation. A violation shall constitute a separate offense with respect to each consumer product involved.

*Criminal Penalties.* Any person who knowingly and willfully violates the Prohibited Acts section of this act after having received notice of noncompliance from the Commission shall be fined not more than $50,000 or be imprisoned for not more than one year, or both.

*Injunctive Enforcement and Seizure.* The United States District Courts shall have jurisdiction to restrain any person from distributing in commerce a product that does not comply with a consumer product safety rule.

*Suits for Damages by Persons Injured.* Any person who shall sustain injury by reason of any knowing violation of a consumer product safety rule may sue any person who knowingly violated any such rule.

*Private Enforcement of Product Safety Rules.* Any interested person may bring an action in any United States District Court to enforce a consumer product safety rule and to obtain appropriate injunctive relief.

## REPAIR

The Sub-Council on Product Safety of the National Business Council for Consumer Affairs recently made these observations on product safety:[1]

The correct installation and repair of a product can often be as important as its safe manufacture. In particular, as a part of installation and repair procedures, steps should be incorporated to assure adequate safety checks on the product. Installation and service personnel employed by retailers need technical training to develop the necessary competence. The

[1]Government Printing Office, *Safety in the Marketplace—A Program for the Improvement of Consumer Product Safety*, April 1973.

use of factory-operated training schools and formal certification of personnel are programs which should be considered to accomplish this end.

Do-it-yourself repairs are a special problem. Instructions on those repairs which are deemed safely accomplishable by normal customers should be clearly described. Conversely an equally clear statement should be made that qualified servicers should be called upon to perform all other repairs.

For the technician, it is wise to use only the replacement parts recommended by the manufacturer for that consumer product. The reason for this is that the manufacturer may have a safety factor in mind that may not be obvious; in fact, by replacing a part with one that seems to have more safety features you may be losing safeguards. In addition to electrical characteristics, there are also physical traits involved. Unsafe replacements can cause overheating and electric shock hazards (through insulation breakdown).

General Electric Corp. supervisor Ray Herzog[2] points out:

Safety depends not only on correct parts but also on installation and wiring of parts. After repairs have been made, a safety check should be performed. Such a safety check includes:

1. Checking lead dress to be sure that no leads or flammable materials touch power resistors, tube envelopes, or other heat-producing components.
2. Making sure that all lead restraints and insulating barriers are properly installed.
3. Measuring continuity between shorted blades of the power cord plug and exposed conductive surfaces (screws, antenna, handle, metal knobs, etc.)

For Step (3) above, the manufacturer's service literature gives acceptable resistance values; these depend on circuit design and the antenna discharge resistor.

Equally important as continuity checking is checking to see that operating high voltage is properly set. Excessive high voltage can cause emission of X-radiation and can lead to an above-average number of failures and possible fire hazards.

[2]"TV Service and Safety," *Electronics World*, February 1971, Copyright 1971 by Ziff-Davis Publishing Company.

# APPENDIX

## FIRST AID

The Occupational Safety and Health Act states:

1. Employer shall insure the ready availability of medical personnel for advice and consultation on matters of plant health.
2. In the absence of an infirmary, clinic, or hospital in near proximity to the workplace which is used for the treatment of all injured employees, a person or persons shall be adequately trained to render first aid. First-aid supplies approved by the consulting physician shall be readily available.

To meet these requirements, which are for your own good, take a standard American Red Cross course and buy a complete first aid kit (not the simple kind found in drugstores) (Figure A-1).

Whenever you wound yourself, take these simple precautions to prevent infection: wash the wound thoroughly with soap and water, apply a recognized antiseptic solution, and then a sterile bandage. If it's a puncture wound, consult your physician—you may need a tetanus booster shot.

For treatment of sprains, strains, bruises, and burns, apply a cold pack to the affected area for 20 to 30 minutes for two or three treatments within the 24 to 36 hour period from the time of the accident (National Safety News, September 1973). To avoid frostbite, keep the temperature of the affected area just above freezing. (In areas where ice or cold water is not readily available, stock the first-aid kit with instant cold packs. These packs contain

**Figure A-1.** Typical industrial first-aid cabinet. (Courtesy of Johnson & Johnson Co.)

chemicals that rapidly lower the temperature of the packages when the chemicals are mixed.)

For chemical burns to the eye, immediately irrigate the exposed eye with large amounts of low-pressure water. Whenever any foreign object is imbedded in the eye, consult a physician.

To treat victims of severe electric shock, refer to Chapter 2.

## LIFTING

When lifting and moving electronic gear, whether it is a color TV console or a rack-mounted power supply, the idea is to save your back and to avoid hernias. Transistorized circuits have made equipment much lighter in weight and thereby have greatly reduced the risk of muscle strains, but there's still enough heavy stuff in use to warrant proper procedures and equipment when lifting and moving (Figure A-2). Just because thousands of people each year suffer injured lower backs is no reason for you to join their unfortunate numbers.

How much you can safely lift and carry depends not only on your physique but also on how fatigued you are, the climate, the size of the object, etc.

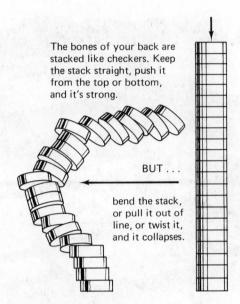

The bones of your back are stacked like checkers. Keep the stack straight, push it from the top or bottom, and it's strong.

BUT . . .

bend the stack, or pull it out of line, or twist it, and it collapses.

**Figure A-2.** Your backbone is like a stack of checkers. (Courtesy of National Safety Council.)

Some state laws and labor union agreements have established definite weight-lifting limits, but these limits are not universal. The International Labor Organization says that an adult male should not normally lift and carry objects weighing more than 82.2 pounds. This is an upper limit and should be scaled down as circumstances demand. (Boys 16 to 18 years old should lift no more than 41.1 pounds.)

One method of avoiding back injuries is to avoid unnecessary lifting:

1. Don't *hand* carry heavy objects—*roll* them, on hand trucks, platform dollies, machine rollers, stair-climbing dollies, etc. Special dollies (see Figures A-3 and A-4) are available for moving awkward items such as TV consoles. If you don't have a loading ramp, build one—pushing a hand truck or dolly up an inclined plane is obviously easier than hand carrying objects up steps.

2. Instead of moving an entire item, use test jigs that will enable you to remove just the chassis or guts and leave the heavy, bulky cabinet behind.

3. Carry your tools and test equipment to the defective equipment, rather than vice versa. Even if the heavy item has already been brought into your shop, it may be wise to put test equipment on a portable cart that can be rolled to the item being repaired.

**Figure A-3.** Proper use of dolly can prevent back injuries. (Courtesy of the Finney Company.)

Whenever you must lift, use the following procedures, which are derived primarily from U.S. Department of Labor recommendations, except as noted:

1. Size up the load first—do not attempt to lift it alone if there is any doubt in your mind about your ability to do so. (Sometimes it is not easy to estimate the weight of a piece of electronic gear. The tendency is to underestimate the weight and to overestimate your ability to lift it. Too often this is discovered in the middle of the lift, when the damage has already been done. When in doubt, therefore, get someone to help you lift and ignore their taunts about your strength.)

2. Place one foot alongside the object being lifted and one behind. Point leading foot in direction of movement. (From National Safety Council's *A New Way to Lift*.)

3. Bend the knees outward and straddle the load somewhat, keeping the back as straight as possible (but not necessarily vertical). Don't stoop.

4. Now start pushing up with your legs, using your strongest set of muscles. Keep the load close to your body (never at arm's length) as you

**Figure A-4.** Finco Handy-Helper Dolly in upright position. (Courtesy of the Finney Company.)

come up, taking full advantage of the mechanical leverage your body now possesses. Lift the object at its center of gravity.

5. Keep your chin tucked in as you lift (per National Safety Council).

6. Lift the object to the carrying position. If it is necessary to change your direction when in the upright position, be careful not to twist the body. Turn your body with changes of foot position.

7. Always have a clear vision over the load.

8. If you deposit the load on a bench or table, place it on the edge to make the table take part of the load and then push it forward with the arms or, if necessary, with part of the body in a forward motion.

9. In putting the load down to the floor surface from a waist-high carrying position, bend the knees and, with a straight back and the load close to the body, lower the load with the arm and leg muscles. In placing your load down on the floor, first be sure that you have

blocks placed to support the load, allowing room to put it down without danger to the fingers.

10. When lifting with a partner, agree beforehand on all your movements so that they can be reasonably synchronized. If this is not done, one of you may lift too soon or shift the load or lower improperly; such actions may overload or strain the partner.

Although improper lifting is the primary cause of lower back injuries, it's possible to get a "bad back" while merely standing or sitting at a workbench. To avoid this, the National Safety Council recommends that you periodically rest one foot off the floor if you're standing for long periods; if you're sitting, they recommend that you raise one or both knees periodically by placing your feet on a foot rest.

Another way to avoid back problems is to make sure your workbench is at the optimum height; if it's too low, you will do unnecessary bending; if it's too high, you will obviously have to keep your hands and arms at an uncomfortable height.

## FIRE PROTECTION

Whether in a large industrial lab or a one-man TV repair shop, wherever there's electronic gear, there is always the chance of fire, unless of course the gear is disconnected from the ac outlet and any outside antennas.

It's much better and easier to prevent a fire than to try to put one out. Here are the steps you should take:

1. Ensure that your electrical wiring conforms to the National Electrical Code—no jury-rig wiring.

2. Protect all antenna circuits from lightning (see Chapter 4.)

3. Fuse every piece of equipment in the shop.

4. Disconnect (at the circuit breaker or switch box) all equipment when the shop is unoccupied. Do not depend on individual equipment power switches.

5. Use soldering irons and torches with utmost care.

6. Install a fire alarm system.

7. If your inventory is expensive, install an automatic sprinkler system.

8. Keep oily rags, waste, and other flammable debris in metal containers with self-closing lids. Do not keep plastic wastebaskets in your shop or lab.

9. Store shipping and packing materials in covered, ventilated metal containers.

10. Keep gasoline out of your shop.

If these techniques fail and fire develops, take these steps:

1. Get everybody out of the area *fast*.
2. Turn in the alarm or call the fire department *fast*.
3. Only after taking steps 1 and 2 should you consider fighting the fire. Too often people have reversed this order and have tried to put out the fire first, only to seriously underestimate its size and the speed with which it spreads.
4. If you decide to fight the fire, make sure you have a clear escape route and that you have the proper fire extinguisher (Figure A-5). For an electronics shop, the proper fire extinguisher is a Type C—Electrical Fires, as odds are that the fire started in live electrical equipment.

**Figure A-5.** Sentry fire extinguisher for Class A, B, C fires. (Courtesy of The Ansul Company.)

To use other types of extinguishers on electrical fires may cause you to be electrocuted or severely shocked.

Under the Occupational Safety and Health Act, fire extinguishers must be conspicuously located where they will be readily accessible and immediately available in the event of fire. They must be located along normal paths of travel, but must not be obstructed or obscured from view.

When the fire is over and firemen have stated that it is safe to reenter the premises, start salvage efforts immediately by removing water, debris, and smoke. A surprising amount of equipment may be saved; indeed, this should be the goal, as insurance may not be sufficient, once depreciation and wear are deducted.

## RF-ENERGY BURNS

RF-energy burns are a particular problem on ships because of the closeness of transmitting antennas to metallic equipment and ship fixtures. On ship, as well as on land, RF burns may be deep and painful and hard to heal. NAVSHIPS manual 0900–005–8000 gives the following precautions:

The handling of metallic cargo lines while shipboard high frequency transmitters are radiating can be hazardous to ship personnel. RF voltages have been detected on crane hooks and other cargo equipment on numerous occasions. These voltages, which can cause injury, are induced in metallic rigging and other metallic items commonly encountered aboard naval ships by nearby transmitter antennas.

An RF burn is the result of RF current flowing through the body area in contact with a source of RF voltage. Any damage that occurs is entirely a result of the heat produced by the current flow through the resistance of the skin at the contacted area. Current flow through a resistance produces heat regardless of the nature of the circuit. The effect of the heat on a person ranges from warmth to severely painful burns. (RF burns should not be confused with RF radiation hazards related to microwave energy.) The criteria used for defining electric shock from power lines are not applicable to the RF burn problem.

The specific level at which contact with RF voltage should be classified as an RF burn hazard is not distinct. *Hazardous* is defined as the RF voltage level that will cause pain, visible skin damage, or involuntary reaction. The term *hazard* does not apply to the lower voltages that cause annoyance, a stinging sensation, or moderate heating of the skin. An open circuit RF voltage exceeding 140 volts on an item in an RF radiation field is to be considered hazardous.

Sometimes it is possible to eliminate RF burn problems aboard ships by relocating the antennas. However, this approach is not usually feasible, since shipboard antennas are usually placed at the optimum locations for desired radiation patterns.

Another approach is to use nonmetallic materials for applications where the hazard of RF burns is a problem. At present there is no suitable nonmetallic substitute for the wire rope used on cargo equipment. However, an insulator link can be used between the crane wire rope and the hook.

The simplest solution is the most obvious one: turn off the transmitter(s) while cargo handling equipment is being used.

## CATHODE-RAY TUBES[1]

Cathode-ray tubes are highly evacuated. These tubes should be handled with extreme care. If a cathode-ray tube is broken, the relatively high external pressure will cause the tube to implode (burst inward), which will result in metal parts and glass fragments being expelled violently as much as twenty feet away.

1. *Handling.* In handling, the tube should be removed and replaced in the packing box with extreme caution (do not drop). The tube face, particularly the rim, must not be struck, scratched, or subjected to more than moderate pressure. When a cathode-ray tube is removed from a unit, it should not be allowed to remain exposed to damage or shock on workbenches, etc., but should be immediately placed in the container provided for that purpose or in the container that held the replacement tube.

2. *Maintenance.* Maintenance or intallation personnel should wear safety glasses and gloves when handling cathode-ray tubes. When removing or inserting tubes in equipment sockets, use only moderate pressure; do not jiggle the tube or stand directly in front of the tube face. Any securing clamp around the rim or face of the tube should be carefully secured.

3. *Disposal.* Defective tubes should be stored or destroyed in such a way that no danger results to personnel.

In addition to the danger involved in implosion, the inner coatings of some tubes are poisonous if absorbed into the blood stream.

[1]Navy Manual NAV MAT P5100.

## — UNDERWRITERS LABORATORIES (UL)

Underwriters Laboratories Inc. is a nation-wide, independent, non-profit testing organization active in public safety since 1894. UL's engineers test products voluntarily submitted by manufacturers to see if the products meet UL's requirements for safety. According to UL, a product is tested and analyzed for all reasonably foreseeable hazards. If the product passes UL's investigation, a UL label is attached to the product. The presence of the label tells you that the product is reasonably free from fire, electric shock, and related accident hazards.

### CABLE TELEVISION

Many of the safety precautions in this book have direct application for CATV technicians, installers, and engineers. Extensive use of aerial lift trucks (see Figures A-6 and A-7) requires extensive safety precautions, which

**Figure A-6.** Barricades and warning signs in place for aerial lift truck. (Courtesy of Multiplier Industries.)

**Figure A-7.** Typical aerial lift truck in operation—Telsta Model SU-34. (Courtesy of General Cable Corporation.)

are given the *CATV Safety Manual*, published by the National Cable Television Association, Inc.

## EMERGENCY LIGHTING

In this age of sudden, unannounced power blackouts, emergency lights that come on automatically when the power goes off are more than just a convenience—they're an absolute must for safety. You cannot count on having matches or flashlights at your finger tips during such emergencies. Without proper light, you may suffer a disastrous fall as you stumble around in the dark. The problem may be even more severe if many people are in the blacked-out area or if the blackout occurs at the same time as another emergency such as fire or storm. Fortunately, many companies manufacture battery-operated lights that operate when commercial power is lost (Figure A-8).

**Figure A-8.** Emergency light.
(Courtesy of Teledyne Big Beam.)

## SIGNS

Although many people ignore warning signs, not every one does. By placing the proper sign at the proper location, you may prevent an accident—signs can announce what is not immediately obvious, especially to the untrained person. Even if a person ignores your sign(s), the sign(s) may help prevent legal suits in case of accident. Appropriate signs are available that are suitable for most hazardous conditions.

## RECOMMENDED READING

In addition to specific publications mentioned in the text, the following standards, reports, and books are recommended for additional information:

*Chapter 2.*
*National Electrical Code*—ANSI.
*Resuscitation Manual*, Edison Electric Institute.
*Chapter 3*
*Static Electricity 1972*, National Fire Protection Association.
*Chapter 4*
*Lightning Protection Code*, National Fire Protection Association.

*Lightning Protection*, by J. L. Marshall, John Wiley & Sons, New York, 1973.
Chapter 6
*Safe Use of Lasers*, American National Standards Institute.
Chapter 10
*Handbook of Noise Measurement*, 7th edition, General Radio.
Chapter 12
*The Consumer Product Safety Act*, Bureau of National Affairs, Inc., Washington, D.C.
*Safety in the Marketplace, A Program for the Improvement of Consumer Product Safety*, Government Printing Office, April 1973.

## ADDRESSES

Chapter 1
National Safety Council
(Publishers of National Safety News)
425 North Michigan Avenue
Chicago, Illinois 60611

Chapter 2
1. American National Red Cross
   Washington, D.C. 20006
2. ECOS Electronics Corporation
   205 West Harrison Street
   Oak Park, Illinois 60304
3. Edison Electric Institute
   90 Park Avenue
   New York, New York 10016
4. Engineering Development Company
   P. O. Box 183
   Warsaw, Indiana 46580
5. Essex International, Inc.
   3501 West Addison Street
   Chicago, Illinois 60618
6. Harvey Hubbell, Incorporated
   Bridgeport, Connecticut 06602
7. Institute of Electrical and Electronics Engineers, Inc.
   (Publishers of IEEE Spectrum)
   345 East 47th Street
   New York, New York 10017

8. International Association of Electrical Inspectors
   (Publishers of IAEI News)
   802 Busse Highway
   Park Ridge, Illinois 60068
9. McKinney, Francis M.
   P. O. Box 8479
   Honolulu, HI 96815
10. Millers Falls Company
    Greenfield, Massachusetts 01301
11. Popular Electronics
    1 Park Avenue
    New York, New York 10016
12. Simpson Electric Company
    5200 W. Kinzie Street
    Chicago, Illinois 60644
13. Square D Company
    1601 Mercer Road
    Lexington, Kentucky 40505
14. Systems Research
    P. O. Box 25280
    Portland, Oregon 97225
15. White Rubber Company
    Ravenna, Ohio 44266
16. Daniel Woodhead Company
    3411 Woodhead Drive
    Northbrook, Illinois 60062

*Chapter 3*
1. Custom Materials
   Alpha Industrial Park
   Chelmsford, Massachusetts 01824
2. Stewart R. Browne Mfg. Co., Inc.
   839 Stewart Avenue
   Garden City, New York 11530

*Chapter 4*
1. AVA Electronics Corporation
   408 Long Lane
   Upper Darby, Pennsylvania 19082
2. Dale Electronics
   Box 609
   Columbus, Nebraska 68601

3. Lightning Elimination Associates
   9102 Firestone Blvd
   Downey, California 90241

*Chapter 5*
1. American National Standards Institute
   1430 Broadway
   New York, New York 10018

2. Bureau of Radiological Health
   Department of Health, Education, and Welfare
   Public Health Service
   Food and Drug Administration
   Rockville, Maryland 20852

3. Electronics
   1221 Avenue of the Americas
   New York, New York 10020

4. The Narda Microwave Corporation
   Plainview, L. I., New York 11803

5. Popular Science
   380 Madison Avenue
   New York, New York 10017

6. Victoreen
   10101 Woodland Avenue
   Cleveland, Ohio 44104

*Chapter 6*
1. American Optical Corporation
   Southbridge, Massachusetts 01550

*Chapter 7*
1. W. M. Bashlin Company
   Grove City, Pennsylvania 16127

2. Broadcast Management/Engineering
   274 Madison Avenue
   New York, New York 10016

3. General Cable Corporation
   P. O. Box 666
   Westminster, Colorado 80030

4. Meyer Industries, Inc
   P. O. Box 114
   Red Wing, Minnesota 55066

5. National Cable Television Association, Inc.
   918 Sixteenth Street, N. W.
   Washington, D.C. 20006

*Chapter 8*
1. Contender Corporation
   P. O. Box 297
   Woodland, California 95695
2. Davis Manufacturing Co.
   1500 South McLean Blvd.
   Wichita, Kansas 67213

*Chapter 9*
1. Miller-Stephenson Chemical Co., Inc.
   Route 7
   Danbury, Connecticut 06813
2. Over-Lowe Company
   2767 South Tejon Street
   Englewood, Colorado 80110
3. Sprayon Products, Inc.
   26300 Fargo Avenue
   Bedford Heights, Ohio 44146

*Chapter 10*
1. Bausch & Lomb
   465 Paul Road
   Rochester, New York 14602
2. H. E. Douglass Engineering Sales Company
   P. O. Box 7209
   Burbank, California 91505
3. Environmental Protection Agency
   1835 K Street, NW
   Washington, D. C. 20460
4. General Radio
   300 Baker Avenue
   Concord, Massachusetts 01742
5. Koss Corporation
   4129 N. Port Washington Avenue
   Milwaukee, Wisconsin 53212
6. Scott Instrument Laboratories
   30 Cross Street
   Cambridge, Massachusetts 02139

7. Sigma Engineering
   Norton Company
   11320 Burbank Boulevard
   North Hollywood, California 91601

8. Tull Environmental Systems
   P. O. Box 4628
   Atlanta, Georgia 30302

*Chapter 11*

1. Channellock, Inc.
   Meadville, Pennsylvania 16335

2. Edsyn, Inc.
   15954 Arminta Street
   Van Nuys, California 91406

3. The Power Tool Institute, Inc.
   Box 1406
   Evanston, Illinois 60204

4. Microflame, Inc.
   3724 Oregon Avenue South
   Minneapolis, Minnesota 55426

5. Xcelite, Inc.
   Thorne and Bank Street
   Orchard Park, New York 14127

*Chapter 12*

1. Consumer Product Safety Commission
   1750 K Street, NW
   Washington, D.C. 20207

*Appendix*

1. The Ansul Company
   Marinette, Wisconsin 54143

2. Finney Company
   34 West Interstate
   Bedford, Ohio 44146

3. Johnson & Johnson
   New Brunswick, New Jersey 08903

4. National Fire Protection Association
   60 Batterymarch Street
   Boston, Massachusetts 02110

5. Stonehouse Signs, Inc.
   P. O. Box 546
   Arvada, Colorado 80001

6. Teledyne Big Beam
   290 East Prairie Street
   Crystal Lake, Illinois 60014

7. Underwriters Laboratories Inc.
   207 East Ohio Street
   Chicago, Illinois 60611

8. Multiplier Industries
   224 North Fifth Avenue
   Mount Vernon, New York 10550

# INDEX

# A

Addresses, 256-61
Aerosols, 189-90
Antennas:
  installation, 178
  lightning protection for, 81-88

# B

Batteries, 190-92
Body resistance, 11-12
Bonding:
  for lightning protection, 79
  for static electricity protection, 58
Broadcast and communications
    transmitters, radiation hazards,
    134–37

# C

Cable laying, 181-82
Cable television, 253 (*see also*
  CATV)
Cadmium, 192
Cardiopulmonary resuscitation,
  45-49
Cathode-ray tubes, 252
CATV, lightning protection, 97 (*see
  also* Cable television)
Chemicals, 188-95
  aerosols, 189-90
  batteries, 190-92
  cadmium, 192
  in hazardous locations, 192
  liquid crystals, 192
  ozone, 192
  in printed circuits, 193
  subsurface gases, 193-94
  sulfur hexafluoride, 195
  trichloroethylene, 194
  in waveguides, 194-95

Chisels and punches, 228
Coaxial lines, lightning protection, 88
Cold-cathode gas discharge tubes, radiation hazard, 118

**D**

Dosimeters, noise, 207-9
Drills, 228

**E**

Ear protection, 217-21
Electricity, static (*see* Static electricity)
Electric shock hazards, 14-18
  testing for, 18
Electric shock statistics, 9-10
Electrometers, 56
Electroscope, gold leaf, 55
Electrostatic voltmeters, 56
Enclosures, shock protection, 40
Eye protection (*see* Lasers)

**F**

Files, 228
Fire protection, 249-51
First aid, 244-45
Floor mat, insulated, 25
Fluorescence-effect tubes, radiation hazard, 118

**G**

Gas-discharge tubes, radiation hazard, 118
Gases, subsurface, 193-94
GFI (*see* Ground fault interrupters)
Gloves, rubber, 24
Glow tubes, neon, 56
Goggles, laser, 152
Grinding wheel, 228
Ground:
  equipment, 26
  ideal, 28
  resistance testing, 29
  system, 26
  unintentional, 27
Ground fault interrupters, 32-38
Grounding, 26-38
  for lightning protection, 78-79
  for static electricity protection, 58-61
Grounding rod, 27
Grounding system defects, 29

**H**

Heart massage (*see* Cardiopulmonary resuscitation)
Heat-effect tubes, radiation hazards, 118

**I**

Injury statistics, 1
Insulation, 22
  double, 23

Insulation checks, 25
Isolation transformers (*see* Transformers, isolation)


# K


Klystrons, radiation hazards, 114-18
Knives, 229


# L


Ladders, 163-72
  climbing, 171-72
  maintenance, 172
  purchase, 167-68
  setup, 168-71
Lasers, 138-62
  biological effects, 143-48
  catastrophic failure, 154
  cryogenic materials, 153
  environmental control, 154-55
  exposure limits (table), 140-41
  goggles, 152
  performance standard, 155-62
  properties, 142
  safety precautions, 148-55
  ultraviolet and x-radiation, 153
Leakage current, 18, 20-22
Let-go currents, 12
Lifting, 245-49
Lighting, emergency, 254
Lightning, 69-97
  personal safety rules, 71, 73, 75
  protective measures
    antennas, 81-88
    bonding, 79

Lightning *cont.*
  protective measures *cont.*
    CATV systems, 97
    coaxial lines and waveguides, 88
    current interruption, 81
    grounding, 78
    land communication facilities, 90-94
    power service and utilization equipment, 94
    station building, personnel, and equipment, 88-90
    voltage limitation, 80
Liquid crystals, 192
Lockouts, 38
Low-voltage circuits, 41


# M


Magnetic-effect tubes, radiation hazards, 118
Microwave ovens, radiation hazards, 120-28


# N


Noise, 196-222
  control, 209-17
  dosimeters, 207-9
  ear protection, 217-21
  effect on man, 201-4
  laws, 204-7
  measurement, 207-9
  sonar, 221-22
  ultrasonics, 221

# O

Occupational Safety and Health Act, 2-5
Occupational Safety and Health Administration, 3-5
OSHA (*see* Occupational Safety and Health Administration)
Ozone, 192

# P

Pliers, side-cutting, 230-31
Pole climbing, 175-77
Powder-actuated tools, 230
Power lines, fallen, 43–45
Power tools, 232-34
Printed circuits, 193
Product safety, 235-43

# R

Radar, radiation hazards, 128-34
Radiation:
    biological effects, 101-6
    electroexplosive devices, effect on, 107-9
    ionizing, effects of, 102-4
    maximum recommended levels (table), 101
    nonionizing, effects of, 105-6
    sources of ionizing (table), 99
    sources of microwave (table), 99
    sources of radio and lf (table), 99
Radiation Control for Health and Safety Act, 100

Radiation hazards, 98-137
    broadcast and communications transmitters, 134-37
    cold-cathode gas discharge tubes, 118
    color TV, 109–14
    fluorescence-effect tubes, 118
    heat-effect tubes, 118
    hydrogen thyratrons, 114
    klystrons, 114-18
    magnetic-effect tubes, 118
    microwave ovens, 120-28
    radar, 128-34
Reading, recommended, 255-56
RF burns, 251-52

# S

Safety concepts, 6
Safety rules, 5
Screwdrivers, 230
Shorting rod (*see* Gounding rod)
Signs, 255
Snips, metal cutting, 229
Soldering iron, 231
Sonar, 221-22
Staple guns, 231
Star drills, 231
Static bar, electronic, 63
Static comb, 61
Static electricity, 50-68
    control, 57
        bonding and grounding, 58
        in hospital operating rooms, 64
        humidification, 63
        ionization, 61
        munitions and explosives, 64
        in semiconductor manufacturing and handling, 64
    detection and measurement, 55-57

Static electricity *cont.*
  generation, 53-55
Static eliminators, radioactive, 63
Static meter, 57
Sulfur hexafluoride, 195

Towers *cont.*
  maintenance, 173
Transformers, isolation, 38
Trichloroethylene, 194
TV, color, radiation hazard, 109

# T

Thyratrons, hydrogen, radiation
    hazards, 114
Tools, 223-34
  chisels and punches, 228
  drills, 228
  files, 228
  grinding wheel, 228
  knives, 229
  powder-actuating, 230
  power, 232-34
  screwdrivers, 230
  side-cutting pliers, 230-31
  snips, 229
  soldering iron, 231
  staple guns, 231
  star drills, 231
  welding, 232
  welding torch, 229-30
  wrenches, 232
Towers, 172-75
  climbing, 174
  erection, 172-73
  lowering, 175

# U

Ultrasonics, 221
Underground hazards, 179-87
    compacting auger, 182-85
    line-laying machines, 185-87
Underwriters Laboratories, 253

# V

Van de Graaff generators, 53-55

# W

Waveguides, 194-95
Welding, 232
Welding torch, gas, 229-30
Workmen's Compensation, 2
Wrenches, 232